COOKIN' WITH HOME STORAGE

All rights reserved. No part of this book may be reproduced in any form without permission from the author.

Cookin' with Home Storage
© Copyright 1998

By Peggy Layton

To obtain extra copies of this and other books on the subject of food storage, see order form in the back of this book or call the phone number below. Wholesale & quantity discounts for stores, churches and other groups available.

Peggy Layton
P.O. Box 44
Manti, Utah 84642
435-835-0311
Fax: 435-835-0312
E-mail splayton@sisna.com
Web pages:
htttp://www.ut-biz.com/homestoragecookin'/
http://www1.icserv.net/D100001/X100043/books.html

About The Authors

Peggy Layton has a Bachelor of Science degree from Brigham Young University in Home Economics Education with a minor in Food Science and Nutrition. Peggy is the wife of Scott R. Layton, and they have seven children. With nine people to feed every day , Peggy uses bulk foods to make the family meals. She saves alot of money and feels good knowing that her family is eating better and getting less fat in their diets. Peggy is currently the Home Economist for Country Harvest. Peggy Layton and Country Harvest, a Winning team, working together to bring you the finest line of products and information ever established in the food storage industry. Look for future cookbooks written by Peggy Layton, using bulk foods from Country Harvest.

Vicki Tate has spent many years on preparedness and helping many families put together their food storage programs. She has a BA degree from BYU in International Relations and a degree in Elementary Education. She organized the Manti, Utah Stake Food Storage Program. She is the wife of Steven Tate and the mother of four children.

Through the years countless people have asked, " What do I do with all this Food storage". That is what prompted the compiling of this cookbook. This cookbook is a collection of recipes which have been compiled from various sources. Many of the recipes have been adapted to cooking with bulk food storage products.

Artwork by Kristine Alder of Manti, Utah

All rights reserved. This book may not be reproduced in any form without permission from the authors.

Peggy Layton
P.O. Box 44
Manti, UT 84642
Phone Orders:(435) 835-0311
Fax # (435) 835-0312

Table of Contents

Historical Overview

In June of 1849, Chief Walker and his brothers asked Brigham Young to send a group of saints to the Sanpitch Valley to settle and teach their people the white man's ways. In August, an exploring party was sent south to determine the advisability of such a settlement. Here the Indians showed them water flowing from each canyon, fertile sagebrush soil, timbered mountains and much tall grass growing on swamplands.

After this favorable report, Brigham Young, in the October General Conference, called 50 men with their families, under the leadership of Patriarch Isaac Morley, to settle the area. They left Salt Lake on October 28, 1849 for the unknown valley by way of Salt Creek Canyon. At this time, Provo was the only community south of Salt Lake City. When they arrived at the present site of Manti, it was November 19, 1849, well into the worst winter the area had seen for many years.

Because of the lateness of the season, Isaac Morley advised the settlers to move to the south side of the grey hill that jutted down into the valley where the Manti Temple now stands. There, some dug caves into the hillside while others made dugouts. Those who still encamped on the creek set their wagon boxes up endwise with their wagon covers stretched across to form an enclosure. This did not work well and soon these saints joined the rest of the group digging into the hillside for shelter.

The snow came. The cattle at first subsisted on the remaining dry grass of the Sanpitch Swamp that the Indians had burned off just before the colonists arrived. Now the colonists were forced, by the deepening snows, to move their cattle south some two miles to Warm Springs where, in despiration, the men and boys worked day after day with shovels winrowing the snow from the grass for the sustenance of the poor starving beasts. The cattle's horns were sharpened by filing to help protect themselves from the ravenous attacks of wolves and coyotes and to help them dig through the snow for food. However, more than half of the cattle died and even though they were frozen, the Indians utilized all the carcasses for food. Of the 240 head of cattle brought to the valley, there were only 100 survivors in June of 1850.

Supplies were running low and word was sent north to Salt Lake for additional food and other necessary items. During the winter and into the spring the colonist's food consisted of scant rations of grain, beans, some milk, a small amount of meat and an occasional egg. From these they made breads, gravy, mush and beans. Their diet was very simple and included many of the authentic pioneer recipes we have included in this book.

When spring came, the settlers were able to pick greens of dandelions, wild mustard, and pig weed which grew on the hills and in the valley. These added much needed vitamins to their scant diet and were a very welcome addition to their food supplies. Warm drinks such as Brigham Tea and peppermint tea were made from wild plants that grew along the ditch banks and in the mountains. These, together with other herbs were put to good medicinal use.

The settlers planted their gardens that year from the seeds they brought with them. As the carrots, potatoes, cabbage, onion, and squash were harvested they were stored in root cellars to help preserve them throughout the winter. The addition of these vegetables made possible a much healthier diet and increased markedly the types of foods the settlers were able to cook; one of the mainstays being hearty vegetable soups. During the summertime they picked wild berries from fields and mountains. Milk gravy, some meat, soups, various breads, beans, porridge, puddings and berry pies were the majority of the foods the settlers were now able to prepare.

In 1853, a group of Danish immigrants were sent south to Manti and the Sanpete area, which brought a strong Scandanavian influence to the area. One of the recipes they brought with them was their famous Danish dumplings.

By the late 1860's and early 1870's fruit trees, purchased at the area greenhouse at Parowan, Utah, were beginning to bear fruit. Apples were particularly popular in their repertoire of recipes because they stored so well.

As the settlement was established stores, meat markets, confectionaries, and the like sprang up. After the valley was settled and well into the next 100 years, the family cow, chickens, pigs and gardens were still the backbone of the Sanpete family diet, as was the case in the other rural communities. It wasn't until World War II that refrigerators, as we know them, became generally available. That, coupled with new marketing and packaging techniques and better transportation, began to change the way people ate. They began to use more sugar, canned goods, a variety of meats and packaged foods. Most gave up the traditional cow and chickens and found it easier to use the quick and easy foods that became so readily available. Sad to say, many of the old skills and much of the ability to create meals out of the very basics has been lost.

As families store commodities for their "home storage" one of the worthwhile things to look at is how the early settlers and "old timers" of the area lived, because so many of our basic storage items are the very foods they ate. Their recipes are invaluable to us. We have many wonderful storage items available today ranging from the very basics, to a wide variety of dehydrated food products and canned goods. It is easy to build a good home storage program, but problems come when we start to use these foods because we are so used to other ways of eating. It is hard to think of meals to create with the foods at hand, or even know how to use some products.

For many years we've seen a need for a comprehensive home storage cookbook to help solve this problem. It is our hope that this book will provide recipes and suggest ways you can use your storage no matter which level it is on.

Hints, Substitutions
and Reconstituting

General Preparedness Hints

1. Make sure you store a good variety of foods of high quality.

2. Use your storage foods so that you are use to cooking with them and your family is use to eating them. This is not something you want to have to learn under stress. It takes about 3 months for your body to adjust to new ways of eating and about as long to adjust psychologically.

3. Make sure you have a hand grinder in your storage.

4. Cooking oil is an extremely important item to have in your storage program. It adds calories and flavor and it is very hard to cook without. Oil makes a great barter item.

5. A good variety of spices and flavorings are an absolute must in a livable storage program.

6. Whenever you serve a bean dish include rice, wheat, or corn as part of the menu. The combination of one of these grains along with the beans provides a complete protein. Some suggestions might be beans served over rice, chili and corn bread, or bean soup and wheat bread.

7. Add a little sugar to your dehydrated vegetables to improve the flavor.

8. To improve the flavor of powdered milk, add a small amount of vanilla and chill.

9. Store your foods in a cool, dry place away from the sunlight to protect it's shelf life.

10. Rotate your storage.

11. Remember - food is only one part of a total self-sufficiency program. Don't forget to add paper products, detergents, medical supplies, bedding, seeds, tools, fuels, and emergency kits.

12. Most of us store four basic items for survival: wheat, milk, honey and salt. There are serious problems associated with this type of diet, such as : Many people are allergic to wheat. Most of them will not even be aware of this until they are trying to live on it meal after meal. Young children can not tolerate large amounts of wheat. Store a variety of other grains such as rye, millet, barley. Too many of us store these foods away thinking we will live on them if we have to. A crisis period is not the time to learn to cook with these items. Appetite fatigue can develop, a person would rather die than eat the same food over & over. Small children and older people are at the most risk.

13.. Also store a variety of beans. This will add variety of color, texture and flavor. It is better to have a 3 month supply of a variety of foods than a years supply of the basics.

14. It is very important that you make sure you have baking powder, soda, yeast, powdered eggs, bouillon, tomatoes, cheese and onions. You cannot cook even the most basic recipes without these items. Vitamins are also important, especially Vitamin C.

15. Add meats to your storage by buying MRE's most of which have meat in them. Also buy a variety of canned meats and fish.

16. Store what you eat & eat what you store.

17. Store your foods in a cool, dry place out of sunlight as these are the things that have the greatest affect on shelf life. If your are using plastic buckets make sure they are lined with a food grade plastic liner. Never use trash liners as these are treated wtih pesticides. A better container is your #10 cans which most of the preparedness compines use when they package their foods.

18. If your stored water has sat for a long period of time, it goes "flat". To improve its flavor aerate it by pouring it back and forth between containers or beating it with a hand beater. It is also a good idea to store flavored drink mixes which can be used to make its flavor more agreeable.

19. To improve the flavor of beef TVP, add one to two tablespoons of worchestershire sauce to one cup of the TVP.

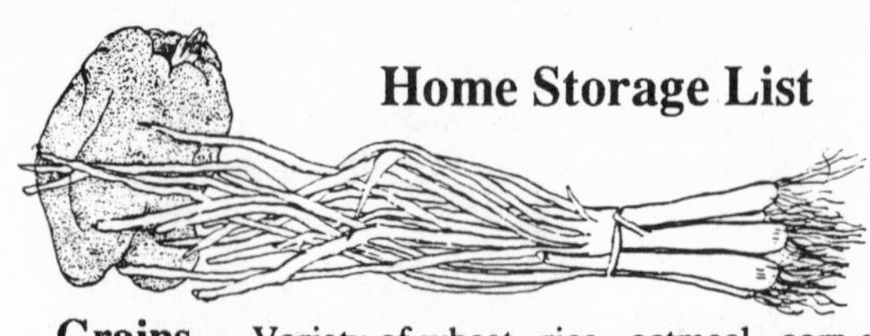

Home Storage List

Grains----Variety of wheat, rice, oatmeal, corn or cornmeal, (field or popcorn) or both, farina, germade, millet, 6-grain, 9-grain, white flour, pasta, etc.

Legumes----Variety of pinto beans, red beans, Anasazi beans, kidney, lentils, sprouting peas, etc.

Sweetners----Honey, white sugar, brown sugar, powdered sugar, molasses, corn syrup, etc.

Dairy Products----Milk and powdered eggs. Powdered margarine is also available.

Salt and a variety of your favorite spices.

Vegatables---(1/2 should be potatoes).(Dehydrated vegetables are a top priority), such as soup and stew blends, carrots, celery, cabbage, peas, green beans, broccoli, beets. Canned and bottled vegetables.

Fruits---(Dehydrated fruits are also a top priority) fruit coctail, apples, applesauce, flavored apples, bananas, peaches, apricots, raisins, etc. Canned and bottled fruit.

Flavorings----Bouillon (beef and chicken), tomato powder, tomato sauce, or canned tomatoes, cheese powder, onions. Baking cocoa, and green peppers are also nice to have.

Cooking Agents----Oil is very important, baking powder, baking soda and yeast.

Seeds for sprouting----Alfalfa, mung, radishes, sprouting peas, lentils, etc.

Vitamins----Especially multi vitamins and vitamin C.

Baby Formula and Food----Where applicable.

Quick and easy to prepare foods----(For times when you are ill or can't psychologically handle cooking with the basics), MRE'S (meals ready to eat), freeze dried casseroles, canned chili, canned soups, canned meats, peanut butter, etc., are nice to have.

Psychological Foods----Puddings, jello, chocolate chips, dream whip, cake mixes, hard candies, powdered drinks, and anything else your family enjoys eating that stores well.
(Store some of these as you begin your storage. They are important. Add to them as you can afford to and in the quantities your family needs.

****Remember**----** It is better to have a 3 month supply of a variety of foods than a years supply of only wheat and beans.

Emergency Substitutions

Item	Amount	Substitute
Baking powder	1 t.	1/4 t. soda and 1/2 t. cream of tarter
Butter or margarine	1 C.	1 1/2 C. margarine or butter powder
Chocolate (Unsweet.)	1 square	3 T. cocoa plus 1 T. butter
Cornstarch	1 1/2 t.	1 T. flour
Corn syrup	1 1/2 C.	1 C. sugar plus 1/2 C. liquid
Egg (Whole)	1 whole egg	2 egg yolks plus 1 T. water or 2 T. dehydrated eggs plus 2 1/2 T. water
Green pepper	1 medium	1/4 C. dehydrated green peppers
Milk, whole	1 C.	1/2 C. evaporated milk plus 1/2 C. water or 1 C. reconst. milk plus 2 T. butter
Onion	1 medium	1/4 C. dehydrated onion
Shortening or butter	1 C.	2/3 C. vegetable oil
Sour cream	1 C.	1 C. milk plus 1 1/2 T. vinegar
Sugar	1 C.	3/4 C. honey (reduce liquid by 1/4 C. or add 1/4 C. flour)
		1 C. molasses
		1 1/2 C. carob syrup
		1 1/4 C. malt syrup
Sugar, powdered	1 C.	1 C. sugar and 1/2 t. cornstarch Blend in blender until powdered.
White flour	1 C.	3/4 C. whole wheat flour
		7/8 C. rice flour
		1 C. corn flour
		1 C. corn meal
		1 1/2 C. rolled
		3/4 C. buckwheat
		1/2 C. barley flour
		3/4 C. rye flour
Yeast	1 package	1 T. yeast

Hints and Substitutions

Equivalent Measurements

Item	Equivalent
Bread	4 slices = 1 cup crumbs
Butter	1/4 pound. = 1/2 cup
Cheese	1 pound = 4-5 cups grated
Flour, sifted	1 pound = 4 cups
Graham crackers	14 squares = 1 cup crumbs
Macaroni, uncooked	4 oz. = 2 1/4 cooked
Rice	1 pound = 2 1/3 cups
Spaghetti, uncooked	7 oz. = 4 cups
Sugar	1 pound = 2 cups
Sugar, brown	1 pound = 2 1/4 cups
Sugar, powdered	1 pound = 4 cups

3 t. = 1 T.	2 cups = 1 pint
2 T. = 1/8 cup	2 pints = 1 quart
4 T. = 1/4 cup	4 quarts = 1 gallon
5 1/3 T. = 1/3 cup	2 gallons = 1 peck
8 T. = 1/2 cup	8 quarts = 1 peck
16 T. = 1 cup	4 pecks = 1 bushel
2 oz. = 1/4 cup	
4 oz. = 1/2 cup	
8 oz. = 1 cup	

Honey Substitution
(Conversion Chart for Recipes)

Sugar	Honey	Subtract Liquid	OR	Add Flour	Plus Soda
1 C.	3/4 C.	minus 1/4 C.	or	plus 4 T.	plus 1/4 t.
1/2 C.	6 T.	minus 2 T.	or	plus 2 T.	plus 1/8 t.
1/3 C.	1/4 C.	minus 1 1/2 T.	or	plus 1 1/2 T.	plus 1/12 t.
1/4 C.	3 T.	minus 1 T.	or	plus 1 T.	plus 1/16 t.

Hint: Cook cakes and other baked goods made with honey on lower temperature.

Hint: Honey will soften cookie batters. If you want the crisp variety of cookies, add 4 tablespoons flour for each 3/4 cup honey used.

Suggested Amounts to Store for One Year

Item	Adult		Children (Ages)				
	Male	Female	1 to 3	4 to 6	7 to 9	10 to 12	13 to 15
Wheat	200 lbs.	150 lbs.	60 lbs.	100 lbs.	150 lbs.	190 lbs.	200 lbs.
Variety of Other Grains	150 lbs.	125 lbs.	50 lbs.	60 lbs.	70 lbs.	90 lbs.	125 lbs.
Variety of Legumes	75 lbs.	50 lbs.	15 lbs.	25 lbs.	50 lbs.	60 lbs.	75 lbs.
Sweeteners	65 lbs.	60 lbs.	40 lbs.	40 lbs.	50 lbs.	60 lbs.	65 lbs.
Powdered Milk	60lbs.	60 lbs.	80 lbs.	80 lbs.	75 lbs.	75 lbs.	75 lbs.
Eggs*	2 cans	2 cans	1 cans	1 can	1 can	2 cans	2 cans
Salt	10 lbs.	10 lbs.	2 lbs.	5 lbs.	5 lbs.	10 lbs.	10 lbs.
Variety of Fruits* (Dehydrated)	8 cans or 25-30 lbs.	8 cans or 25-30 lbs.	4 cans or 15 lbs.	4 cans or 15 lbs.	6 cans or 20-25 lbs.	6 cans or 20-25 lbs.	8 cans or 25-30 lbs.
Variety of Vegetables* (Dehydrated)	15 cans or 40-45 lbs.	15 cans or 40-45 lbs.	6 cans or 15-18 lbs.	8 cans or 20-25 lbs.	10 cans or 25-30 lbs.	12 cans or 30-35 lbs.	15 cans or 40-45 lbs.
Bouillon*	1 can	1 can	1/2 can	1/2 can	1 can	1 can	1 can
Onions*	1 can	1 can	1/2 can	1/2 can	1 can	1 can	1 can
Cheese*	2 cans	2 cans	1/2 can	1 can	2 cans	2 cans	2 cans
Tomato Powder	2 cans	2 cans	1/2 can	1 can	2 cans	2 cans	2 cans
Oil	2 gal.	2 gal.	1 gal.	1 gal.	2 gal.	2 gal.	2 gal.
Yeast	2 lbs.	2 lbs.	1 lb.	1 lb.	2 lbs.	2 lbs.	2 lbs.
Sprouting Seeds	10 lbs.	10 lbs.	2 lbs.	4 lbs.	6 lbs.	10 lbs.	10 lbs.

* Cans are #10 sized cans

Dried Food	Amount	Water	Yield
Apple Granules	1 cup	3 cups	3 cups
Apple Slices	1 cup	1 1/2 cups	2 cups
Apricot Slices	1 cup	2 cups	1 1/2 cups
Beets	1 cup	3 cups	2 1/2 cups
Bell Peppers	1 cup	1 1/2 cups	2 cups
Buttermilk	1 cup	1 1/2 cups	2 1/2 cups
Cabbage	1 cup	2 1/2 cups	2 cups
Carrots	1 cup	2 cups	2 cups
Celery	1 cup	1 cup	2 cups
Cheese	1 cup	1/3 cup	2/3 cup
Corn (Sweet)	1 cup	3 cups	2 cups
Dates	1 cup	1 cup	1 1/3 cups
Fruit Blend	1 cup	1 1/2 cups	1 1/2 cups
Gelatin	1 cup	4 cups	4 cups
Green Beans	1 cup	2 cups	2 cups
Margarine	1 cup	2 tablespoons	3/4 cup
Milk	1 cup	4 cups	4 cups
Onions	1 cup	1 cups	1 1/2 cups
Onions (Minced)	1 cup	1 1/2 cups	2 cups
Peach Slices	1 cup	2 cups	2 cups
Peanut Butter	5 tablespoons	4 t. oil + 1/3 t. salt	1/2 cup
Peas	1 cup	2 1/2 cups	2 1/2 cups
Potato Dices	1 cup	3 cups	2 cups
Potato Granules	1 cup	5 cups	5 cups
Potato Sliced	1 cup	2 cups	1 cup
Sour Cream	1 cup	6 tablespoons	3/4 cup
Spinach Flakes	1 cup	1 1/2 cup	1 cup
Tomato Powder	1 cup	1 1/2 cup	1 3/4 cups

Hint: The amount of water needed can vary according to the moisture content of the product. You may find that you want to increase or decrease the water to suit your needs.

Examples of Reconstitutions

Dried Cabbage

1 3/4 C water **1/2 C dried cabbage**
1/2 t. sugar

Bring water and cabbage to a boil. Reduce heat and simmer 5-10 minutes until tender. Drain the water and add sugar, salt, and pepper to taste. Dot with butter. Fried bacon pieces or corned beef are good added to the cabbage. It is excellent in soups.

Dried Carrots

1 C water **1/2 C carrots**
1/2 t. sugar

Soak sliced carrots for 10 minutes before cooking. Simmer 10-20 min. Drain. Add sugar, salt, pepper, and butter.

Dried Green or Red Peppers

1/2 C water **2 T. dried peppers**

Soak in water until rehydrated. Peppers are a wonderful flavoring agent. Add to soups, etc..

Dried Sweet Corn

2 1/2 C water **1 C sweet corn**
1 t. sugar

Soak corn in water for 45 minutes . Bring to a boil. Reduce heat and simmer for 30 more minutes until tender. Drain. Add sugar, salt, and pepper to taste. Dot with butter. Nice added to soups.

Dried Green Beans

3 C water
1 t. sugar
1 C dried green beans

Bring water and beans to a boil. Reduce heat and simmer 20 minutes until tender. Drain. Add sugar, salt, and pepper to taste. Dot with butter. Add1/4 C. onions, bacon bits, or add 1 can of Mushroom soup and green beans and bake at 350° for 20 minutes.
Good added to vegetable soup.

Dried Peas

2 1/2 C water
1/2 t. sugar
1 C dried peas

Bring peas and water to a boil. Reduce heat and simmer 20 minutes until tender. Drain. Add sugar, salt, and pepper to taste. Dot with butter

Dried Spinach Flakes.

1 C water
1 C dried spinach flakes

Bring water to a boil. Add spinach flakes and simmer 3 minutes. Drain. Salt and pepper to taste. Dot with butter or sprinkle with margarine powder. You may also add 1 t. vinegar.

Dried Tomato Powder

Tomato sauce

1 1/2 C water
1 C tomato powder
1/4 t. sugar

Tomato paste

1 3/4 C water
1 C tomato powder
1/2 t. sugar

Bring water, tomato powder, and sugar to a boil.. Simmer for 10 minutes .

Dried Potatoes

Diced

3 C water
1 C dried diced potatoes
1/2 t. sugar

Sliced

2 C water
1 C dried sliced potatoes
1/2 t. sugar

Bring water to a boil. Simmer 15 minutes until tender. Drain and add sugar, salt, and pepper to taste.

Cooking guidelines for beans

Dry Beans	Cooking time	Pressure Cook (15 lbs)
Black beans	2 hours	5 min
Black-eyed peas	1/2 hour	Do not pressure cook
Great Northern beans	1 1/2 hours	3 min.
Kidney beans	2 hours	3 min.
Lentils	1/2 hour	Do not pressure cook
Lima beans, large	1 hour	3 min.
Pinto beans	2 hours	10 min.
Navy beans	2 hours	7 min.
Split peas	1/2 hour	Do not pressure cook

Hints

1. Rinse all beans and legumes in cold water. Remove all dirt, rocks, or bad beans.

2. Soak the beans in 3 times the amount of water as beans. They can be soaked overnight. Lentils and split peas do not need to be soaked.

3. Quick soaking method: Boil the beans in water for 2 minutes, remove from heat, cover, and let stand for 1 hour.

4. Add 1 t. salt per cup of beans and use a large enough pan because the beans double in volume.

5. Add 1/8 t. baking soda and 1 T. cooking oil to each cup of beans while soaking. This will shorten the cooking time and decrease foaming.

6. Add meat, onions, celery, and herbs during cooking to add more flavor. Add tomatoes, catsup, vinegar and other acid foods after the beans are tender. The acid prevents softening of the beans.

7. Cooked beans freeze well and will keep up to 6 months in the freezer.

Reconstituting Powdered Eggs

Amount of eggs	egg powder	water
1 egg	2 T.	2 1/2 T.
2 eggs	4 T.	5 T.
3 eggs	6 T.	7 1/2 T.
4 eggs	8 T.	10 T.

Unflavored Gelatin

(A Substitute For Eggs)

To substitute for 1 egg : mix 1 t. gelatin, 3 T. cold water, 1/2 cup boiling water. For 2 eggs: mix 2 t. gelatin , 1/3 cup cold water, and 1/2 cup boiling water.

Before starting to mix cookies, cake, or something else : Place cold water in a mixing bowl and sprinkle gelatin in it to soften. Mix thoroughly. Add all the boiling water and stir until dissolved. While preparing the batter, place mixture in the freezer to thicken. When recipe calls for an egg , take the mixture and whip it until it is frothy. Then add it to the batter.

Reconstituting Powdered Milk

For this amount	Mix dry milk		With water
1 quart	Inst. 1 C.	Reg. 3/4 C.	4 C.
1 pint	Inst. 1/2 C.	Reg. 1/3 C.	2 C.
1 cup	Inst. 1/4 C.	Reg. 3 T.	1 C.
1/2 cup	Inst. .2 T.	Reg. 1 1/2 T.	1/2 C.
1/4 cup	Inst. 1 T.	Reg. 3/4 T.	1/4 C.

Substitutions Using Powdered Milk

Whole milk

1 C water
1/3 C powdered milk

Evaporated milk and Whipped Topping

1 C water
2/3 C powdered milk

This milk can be chilled and whipped into a topping by adding 1/2 t. lemon juice. After it is whipped, fold in 1 T. sugar to taste.

Buttermilk

1 C water
1/3 C powdered milk
1 T. vinegar or lemon juice

Let mixture stand in a warm place until thickened (about 18 hours.) Stir until smooth. Refrigerate. A buttermilk freeze dried culture can be purchased at a grocery or health food store, and kept indefinately.

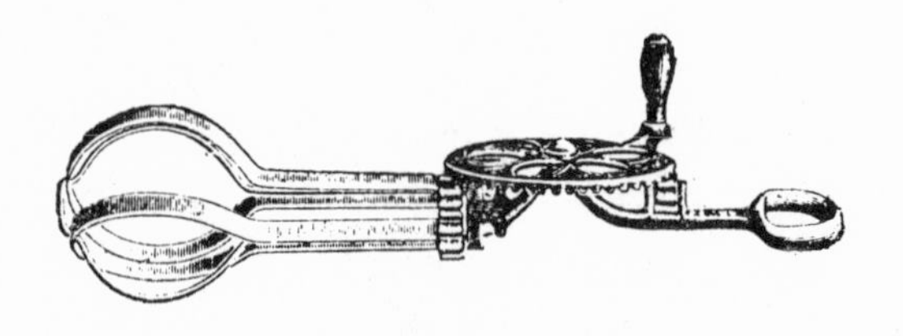

Spices and Condiments
Kristine

Allspice: It has a delicate flavor that resembles a blend of cloves, cinnamon and nutmeg. Uses: pickles, meats, puddings, pies, and drinks.

Anise: The flavor of Anise is that of a sweet licorice taste. Uses: fruits, cakes, rolls, pie fillings, stews, and soups.

Basil: Has an mild, leafy, lemon flavor. Uses: tomato dishes and soups, also in squash and beans and sprinkled over meat.

Bay Leaves: Gives off a pungent, herbal flavor. Uses: vegtetables, stews, seafoods, and soups.

Caraway: Has a flavor of rye bread. Uses: breads, cheese spreads, cookies, vegetables, roast pork.

Cardamon: Comes from the ginger family and has a bitter sweet flavor. Uses: fruit, pastries, cakes, custards, sweet potatos, and pumpkin dishes.

Cayenne: Very Hot. Uses: mexican cookery, chili, beef, stews, cheese souffles, and green vegetables.

Celery Seed: Tastes alot like bitter celery. Uses: dips, soups, slaw, tomatoes, and salad dressings.

Chili Powder: Has a distinctive, hot, spicy flavor. Uses: seafood cocktails, soups, beans, mexican cooking, and cheese sauces.

Chives: Have a mild green onion flavor. Uses: Potatoes, sauces, dips, and salads.

Cinnamon: Has a sweet , spicy flavor. Uses: cakes, cookies, puddings, fruit pies, spiced beverages and pumpkin dishes.

Coriander: From the parsley family, more spicy. Uses: beans, salads, eggs, cheese, pork, sausage, curry sauce, rice and pickles

Cloves: Spicy, sweet, pungent flavor. Uses: ham, apples, pumpkin and mince pies, baked beans, teas, spice cake, and puddings.

Cumin: Salty, balsam like flavor. Uses: cheese spreads, deviled eggs, chicken, dressings, lamb, enchilada sauce, beans, breads and crackers.

Curry Powder: Exotic with heat. Uses: all Indian cooking, chicken, eggs, rice, vegetables and fish.

Dill: Similar to caraway, but milder and sweeter, has a slight bitter flavor. Uses: mostly in pickling, also in salads, soups, dips, and cheeses.

Funugreek: Has a maple flavor, not as sweet. Uses: Indian dishes, candies, cakes, cookies, and oriental cooking.

Garlic: From the onion family, it has a pungent flavor. Uses: dips, soups, vegetables, potatoes, meats, sauces, and bread.

Ginger: Has a fragrant, hot, spicy, sweet flavor. Uses: cookies, cakes, pies, puddings, applesauce, stews, fish and stuffing.

Horseradish: Taste like parsnip, quite hot. Uses: dips, spreads, seafoods, pork, lamb, marinates, and cocktail sauces.

Mace: Similar to nutmeg. Uses: tomato juice, soups, fish, stews, pickling, gingerbread, cakes, welsh rarebit, chocolate dishes and fruit pies.

Marjoram: A delicate herbal flavor. Uses: soups, meats, eggs, sauces, and fish.

Mint: Has a sweet leafy flavor. Uses : jelly, fruit salad, lamb, and tea.

Mustard: A sharp, spicy flavor. Uses: salads, pickling, Chinese hot sauce, cheese sauce, vegetables, molasses cookies, and fish.

Nutmeg: This has a sweet, exotic flavor. Uses: doughnuts, eggnog, custards, spice cake, pumpkin, puddings, and sweet potatoes.

Oregano: A relative of Marjoram, quite a bit stronger. Uses: pizza, spaghetti sauces, meat sauces, soups, vegetables and Italian specialties.

Paprika: A very mild taste, related to the bell pepper. Uses poultry, goulash, vegetables, soups, stews salad dressing, meats, and cream sauces.

Parsley: From the celery family, has a mild flavor. Uses: soups, salads, meat stews, all vegetables, and potatoes.

Pepper: Has a spicy, enduring aftertaste. Uses: most all foods except those with sweet flavors.

Peppermint: A strong minty flavor, quite soothing in tea. Uses: cream cheese spreads, coleslaw, lamb, garnishes, teas, and ices.

Poppy Seeds: A seed that is crunchy, and nutlike. Uses: breads, rolls, cookies, salads and cakes.

Rosemary: Has a delicate, sweetish taste. Uses: lamb dishes, soups, stews, beef, and fishes.

Saffron: Is a very strong, exotic spice, use sparingly. Uses: rice, breads, fish stew, chicken soup, cakes, fish sauces

Sage: Has a strong flavor of camphoraceous and minty. Uses: meat and poultry, stuffings, sausages, meat loaf, hamburgers, stews and salads.

Savory: It has a mild pleasant taste. Uses: scrambled eggs, poultry stuffing, hamburgers, fish, tossed salad, and tomatoes.

Sesame Seeds: Has a crunchy, nutlike flavor. Uses: breads, rolls, cookies, salad, fish, and asparagus.

Tarragon: This herb has a faint anise flavor. Uses: marinates for meats, poultry, omelets, fish, soups, and most vegetables.

Thyme: Has a strong, distinctive flavor. Uses: poultry seasoning, croquettes, fish, eggs, tomato dishes, and vegetables.

Turmeric: This come from the ginger family and has a mild, ginger-pepper flavor. Uses: pickles, salad dressings, rice, and seafoods.

Taco Seasoning Mix

2 t. dried onions
1 t. salt
1/2 t. minced garlic
1/2 t. cornstarch
1/2 t. cumin
1 t. chili powder
1/4 t. oregano

Sloppy Joe Mix

1 T. dried onions
1 t. green pepper flakes
1/2 t. minced garlic
1/4 t. dry mustard
1 t. salt
1 t. cornstarch
1/4 t. celery seeds
1/2 t. chili powder

Chili Seasoning Mix

1 1/2 t. flour
1 T. dried onion flakes
3/4 t. chili powder
1/4 t. dried red pepper
1/4 t. minced garlic
1/2 t. seasoning salt
1/4 t. sugar
1/4 t. cumin

Spaghetti Seasoning Mix

1 T. minced onion
1 T. parsley flakes
1 T. cornstarch
1 T. salt
1/4 t. instant minced garlic
1 t. sugar
1 T. dried green pepper
1/4 t. basil
1/4 t. thyme
1/4 t. sage
1/4 t. marjoram
1/4 t. oregano

Combine all ingredients until mixed well. Store in an airtight container. This is equivalent to 1 pkg. of seasoning mix from the store.

French Dressing Mix

1/8 t. onion powder
1 t. dry mustard
1/4 C sugar
1 1/2 t. salt
1 1/2 t. paprika

1/4 C vinegar
3/4 C vegetable oil

Combine all ingredients. Blend until smooth. Chill before serving

Ranch Dressing Mix

1 T. parsley flakes
1/8 t. garlic powder
2 t. minced onion
1/2 t. MSG
1/2 t. salt

1 C mayonaise
1 C butter milk or sour cream

Combine all ingredients in a bowl. Blend until smooth. This can also be used as a dip for vegetables.

Dry Onion Soup Mix

4 T. Beef Bouillon granules
8 t. dried onion
1 t. onion powder
1/4 t. Bon Appetite season

Chicken Coating Mix

1 t. onion salt
1 T. celery salt
1 T. ginger
1 t. pepper
1 t. sage
1 T. paprika
2 T. parsley flakes
1 T. oregano
1 T. marjoram
1 T. thyme
2 t. rosemary
1 t. garlic

Combine all of the above ingredients. Store in an airtight container. Makes enough coating mix for 1 chicken.

Crispy Coating Mix

1/2 t. pepper
1 t. celery salt
2 t. salt
4 t. parsley flakes
1 C wheat germ
1 t. onion salt
1 t. dry mustard
1 T. paprika
1/2 C sesame seeds
3 C corn or bran flakes

Combine all ingredients. Store in an airtight container. Use this for coating chicken, pork chops, or other meat.

Vegetable Dip

1 C sour cream (reconstituted sour cream powder can be used)
1 C mayonaise
1 T. dill weed
1 T. minced onion
1 T. parsley flakes
1 T. Lawry's seasoned salt

Mix well, chill and serve with vegetables as a dip.

Mayonaise

1 whole egg
2 t. sugar
1/4 t. salt
1 1/2 T. vinegar
dash of pepper
1 C oil

Put all ingredients in the blender, except for 3/4 C. of the oil. Blend together well. While blending slowly add the remaining oil until the mayonaise is thick.

Tomato Sauce

2 C chopped onion
3 cloves garlic
3 T. oil
3 1/2 C bottled tomatoes
2 small cans tomato paste
2 C water
1 bay leaf
1/2 t. salt
1/4 t. pepper
1/4 t. oregano
1/4 t. basil

Saute onion and garlic in oil. Add tomatoes, paste, bay leaf, salt and pepper. Simmer 2 hours. Add more water if necessary. Add oregano and basil. Cook another 15 minutes until thick.

Tomato Sauce

(dried foods)

1 C tomato powder
3 C water
1/2 t. salt
1 t. sugar
1 T. oil
1 T. margarine powder
1/4 C onion flakes
1/4 t. garlic powder
1 bay leaf
pepper to taste

Simmer the tomato powder, water, sugar and salt on low heat for 20 minutes. Saute onions in oil and margarine powder until tender. Add onions and remaining ingredients to the tomatoes and simmer another 15 minutes. Stir often, until thick.

Tomato Catsup

1 cup tomato powder
2 t. sugar
1/4 t. vegetable oil
1/2 t. salt
1/8 t. pepper
Dash of onion powder
1 1/2 C water

Combine all ingredients. Boil, then simmer for 10 minutes until thick.

Thirst Quenchers
-Drinks-
CIDER
SUGAR

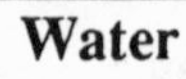

Water

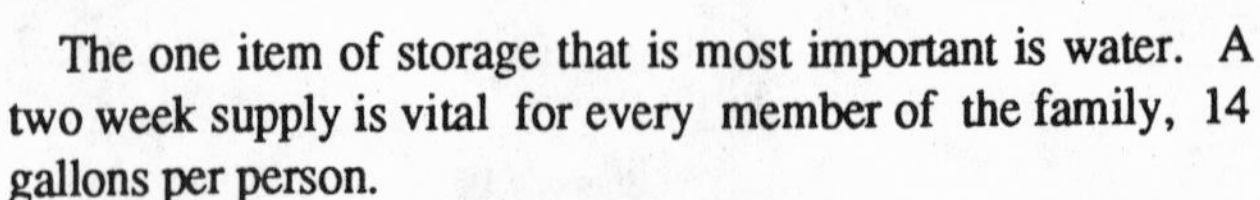

The one item of storage that is most important is water. A two week supply is vital for every member of the family, 14 gallons per person.

If water is treated for bacteria at the time it is stored, and placed in a clean tightly sealed container, and stored away from sunlight, it will be safe for drinking.

If the water you have is cloudy, it must be clarified first before it is sterilized. The way this is done is to add 1/4 teaspoon of powdered alum per gallon of water. The alum reacts with the water to absorb the impurities and settles them to the bottom. After the water has settled, which takes about 30 min. to 1 hour, pour the clean water off into another container. Be sure to discard any cloudy water and all sediment at the bottom. If you don't have any alum, then strain the water through a tightly woven cloth. After all sediment is removed the water is ready to be disinfected.

There are several ways to sterilize or disinfect the water.

1. It can be boiled for fifteen minutes.
2. A solution of household bleach can be added, 1/8 teaspoon per gallon. This is about 5 drops. Mix it up and let stand for at least 30 minutes.
3. Water purifying tablets can be used as to the directions on the bottle. This can be obtained from a pharmacy.
4. 2% tincture of iodine can be used. Add 3 drops of iodine to each quart of water, 6 drops if the water is cloudy. Let this solution stand for at least half an hour to an hour before using. The iodine can be obtained from a pharmacy.
5. Iodine Crystals are most effective in purifying contaminated water. In the crystal form iodine has an infinate shelf life and is ve ry inexpensive. Do not touch the crystals, they can cause burns. Place 4-8 grams in a small glass jar and add water to fill. Shake well. Add 3 teaspoons of solution to a quart of clear water to be treated. Let this stand for 30 minutes. If the water is cloudy, double the amount of solution per quart. When the water is gone, fill it again. The crystals are good for 1,000 quarts of water.

In the early days, root beer was made from dandelion roots, yarrow (gathered wild), and hops. These were steeped in water, then strained. The liquid was placed in a crock, then put behind the stove to ferment for 3-4 days. If they wanted to speed up the process they would add some sugar to the "brew". Ice was added to make it nice and cool. This drink was often enjoyed at parties and always served to the hired hands at hay thrashing.

Brigham Tea
(Pioneer Recipe)

Brigham Tea is readily found growing wild in the mountains in southern Utah. The early pioneers drank it to combat the scurvy that was so prevalent. It is very rich in vitamin C.

Pick the plant. It can either be dried or used fresh. Boil the water desired, take it off the heat, add the herb, and steep for 20 min. Strain it and it's ready to drink. Honey can be added to taste.

Elta Alder

Rosehip Tea
(Pioneer Recipe)

Collect the rosehips from the rose bushes, either wild or domestic, at the end of the summer. Dry the herbs or use them fresh, approximately 1 tablespoon per cup of water. Boil the water desired, take off the heat and add the rose hips and steep 20 minutes. Add honey to taste.

Mormon Sage Tea
(Pioneer Recipe)

Brew the leaves of the sage plants in the same manner as above. Sweeten as desired.

Barley Drink
(Pioneer Recipe)

The barley grain is browned in the oven, then ground into cracked grain, and brewed into a delicious drink sweetened with honey or sugar.

Old Fashioned Egg Nog

2 C. milk
2 eggs
Grated nutmeg
1/2 t. vanilla
2 T. honey or sugar

Blend together. Top with spice and serve. Variation: Add a mashed banana.

Elaine Westmoreland

Country Egg Nog

4 eggs
1/2 C. honey
2 quarts chocolate milk
1/4 C. whipped topping

Add beaten eggs to honey, add chocolate milk and beat. Reserve a few tablespoons whipped topping, fold in remainder. Garnish with reserved topping. Yield 3 quarts.

Cherie Harmon

Hot or Cold Cocoa Mix

7 1/2 C instant powdered milk
1 1/2 C sugar
1 C cocoa
1 1/2 t. salt

Stir 1/2 C mix to 1 C boiling water. Stir to dissolve. Chill if used cold.

Peggy Layton

Milk Shake

3/4 C. water
1/2 C. powdered milk
4 t. flavored syrup (Chocolate, strawberry or your own favorite)

Blend until thick and frosty. Fresh fruit and sugar may be substituted for syrup.

Elaine Harmston

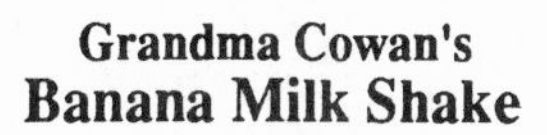

Grandma Cowan's
Banana Milk Shake

1 banana, mashed
2 1/2 C. milk
1 t. sugar

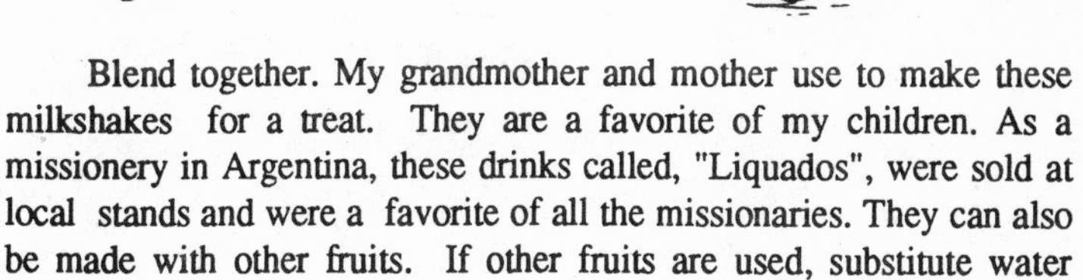

Blend together. My grandmother and mother use to make these milkshakes for a treat. They are a favorite of my children. As a missionery in Argentina, these drinks called, "Liquados", were sold at local stands and were a favorite of all the missionaries. They can also be made with other fruits. If other fruits are used, substitute water for milk.

Vicki Tate

Orange Banana Shake
(Dried or fresh)

2 T Tang or orange breakfast drink
1 banana or (1/2 C reconstituted banana chips)
2 C water
1/3 C pdr. milk + 1 C water
2 T. honey or sugar

Blend until foamy. This makes a refreshing drink.

Peggy Layton

Refreshing Fruit Drink

2 C any bottled fruit and juice
3/4 C milk or (4 T. pdr. milk mixed with 3/4 C water)
1 T. lemon juice

Blend in the blender until smooth . Add 1/2 tray of ice cubes and blend into a slush.

Tangy Orange Drink
(Dried foods)

1 T. Tang or orange drink powder
3 T. pdr. milk
2 eggs
1 C water
dash of nutmeg
sugar to taste

Combine ingredients and blend in a blender. Pour in a glass and sprinkle lightly with nutmeg. Sugar can be added to taste but is not necessary.

Elaine Harmston

Peanut Butter Milk Shake
(Dried foods)

3 C water
1/2 C powdered milk
1/2 C powdered peanut butter powder
1/3 C sugar
crushed banana chips (optional)

Blend all ingredients together using the blender.

Natelie Simmons

Hot Holiday Drink

1/2 C Tang or powdered orange drink
4 C boiling water
1/4 tsp nutmeg
cinnamon sticks or 1/2 t. cinnamom

Add ingredients to boiling water, Stir to dissolve. Serve hot.

Lisa Conn

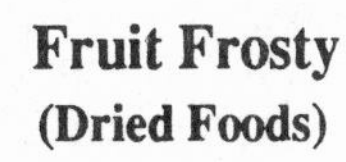

Fruit Frosty
(Dried Foods)

1 C dehydrated fruit (rehydrated in 1/2 C water)
1 C ice
1 C unflavored yogurt
3/4 C milk
3 T. sugar

Blend fruit and water in blender for several minutes. Add other ingredients and blend 30 sec. more.

Vicki Tate

"Mormon Tea" Mix

2 C of Tang or dehydrated orange drink
6 oz pkg. lemonade
1 3/4 C sugar
1 t. cloves
1 t. cinnamon

Mix all ingredients together and store in a covered container. To use: Dissolve 1 T. mix in 1 C. boiling water.

Vicki Tate

V-8 Vegetable Juice

1 bushel, juiced tomatoes
3 med onions
2 yellow squash
2 lg green peppers
12 carrots
10 garlic buds
1 bunch celery
2 zucchini squash
2 large bunches of parsley
4 large bay leaves

Put a string around the 2 bunches of parsley and tie it securely. Put tomato juice into large kettle. Add a little water in another large pan and boil until tender the celery, onions, pepper, carrots, garlic, and squash. Put all these ingredients into a juicer and add to the tomatoes. Add the parsley and bay leaves. Simmer 6 to 8 hours. Put the hot V-8 juice into prepared quart jars. Pressure for 35 min. at 10 lbs. pressure.

Mary Lois Madson

Tomato Juice

1 can tomato sauce
3 cans water
salt to taste

Blend above ingredients and add salt to taste. Hot sauce or tobasco sauce may be added to make a drink like "Snappy Tom" tomato juice.

Elaine Harmston

Ice was often used to cool drinks or food in the ice box. Amazingly, it was available most of the year because of the storage technique used. In the winter, blocks of ice were cut when the river froze over, then hauled to the ice house where it was stacked alternating with layers of straw. This provided such good insulation that it lasted nearly through the summer.

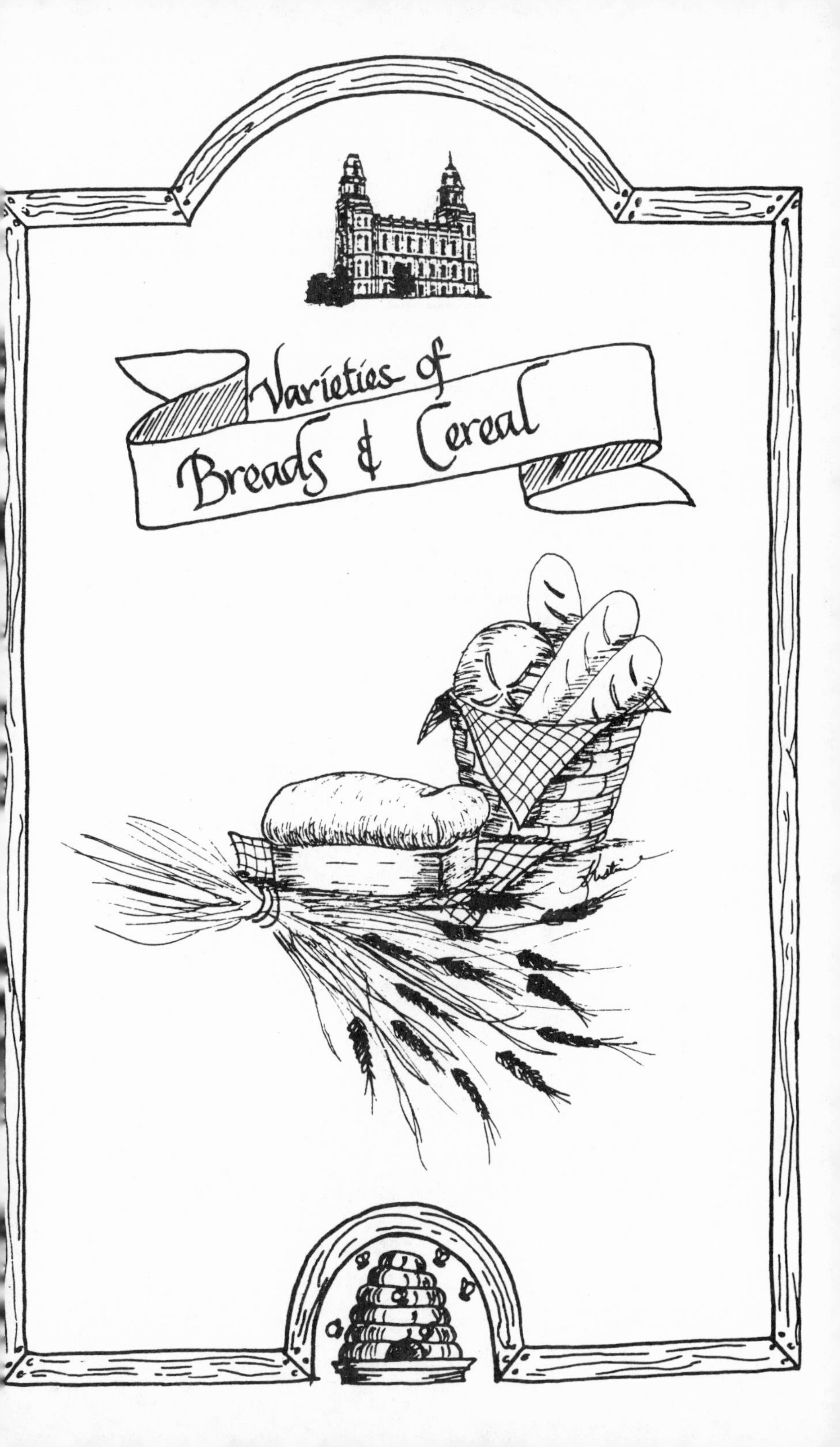
Varieties of
Breads & Cereal

Wheat Flour for White Flour Substitution

1 C. minus 2 T. wheat flour = 1 C. white flour

If using baking powder or soda increase by 1/3. If using yeast, double the yeast.

Hints for Using Wheat

1. If possible, grind wheat just before you use it to retain its full nutrition. It is rich in vitamin E and wheat germ, both of which are soon lost after grinding.

2. Wheat sprouts are good as a snack or in salads. The sprouts add vitamin C and vitamin A to your diet.

3. Uses for wheat:
   a. flour
   b. cereal for breakfast
   c. chili
   d. substitute for potatoes in stew
   e. substitute for rice in fruit salad
   f. Add (cooked) to tossed salad

Homemade Yeast

(Old Recipe)

Liquid yeast: Early in the day, boil **one ounce of best hops** in **two quarts of water** for thirty minutes; strain and let the liquid cool to warmth of new milk; put it in an earthen crock, or bowl. Add **4 tsp. each** of **salt and brown sugar**; now beat up **2 C. of flour** with part of liquid and add to remainder, mixing well together and set aside in warm place for three days, then add **1 C. smooth, mashed boiled potatoes.** Keep near the range in a warm place and stir frequently until it is well fermented; place in a sterilized, wide mouth jug or a glass fruit jar. Seal tightly and keep in a cool place for use. It should thus keep well for two months and be improved with age. Use same quantity as other yeast, but always shake the jug well before pouring out.

Dry yeast cakes: To a quantity of **liquid yeast** add enough **sifted flour** to make a thick batter; stir in **1 tsp. salt** and set to rise. When risen, stir in **sifted and dried cornmeal,** enough to form a thick mush; set in warm place and let rise again; knead well and roll out on a board to about one-half inch thickness and cut into cakes one and one-half inches square or with a two-inch round cutter; dry slowly and thoroughly in warm oven; keep in cool, dry place for use. Will keep fresh for six months. To use, dissolve one cake in **1 C. of lukewarm water.**

Fried Whole Wheat

2 T. oil
1 C. green pepper, diced
2 C. cooked whole kernel wheat
2 T. soy sauce
2 eggs
2 C. coarsely chopped onion
2 C. left-over meat (bacon, ham, shrimp, chicken)
1/2 tsp. salt

In a large skillet, saute onion and green pepper in oil. Then add meat and whole wheat. Add 2 eggs, beaten lightly, soy sauce, and salt. Stir until eggs are cooked.

Boston Baked Whole Wheat

4 C. whole kernel wheat
1 lb. bacon, cut in fourths
1/4 C. molasses
1/3 C. catsup
1/2 tsp. dry mustard
10 C. water
1 large onion, diced
1/3 tsp. pepper
2 tsp. salt

In a large roaster or Dutch oven, combine wheat, water, bacon, and onion. Combine remaining ingredients in bowl and pour into pan with wheat. Cover and bake at 200° for 6 hours. Remove cover the last 1/2 hour of baking. Add a little boiling water if mixture becomes a little dry. Serve hot with bread. Makes 14 Cups.

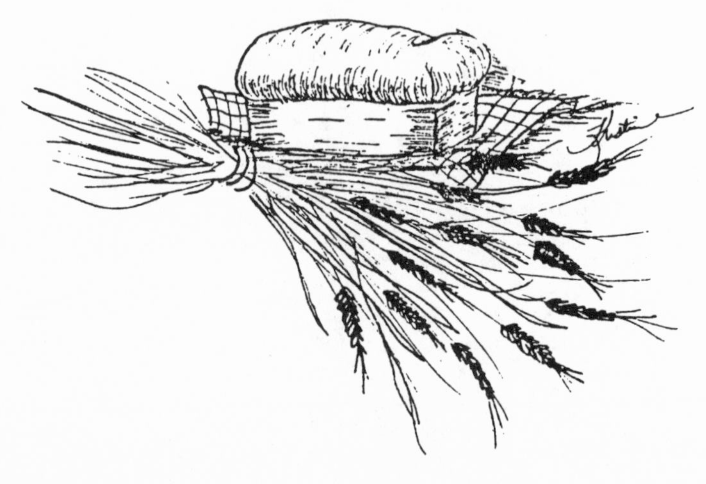

One prized item brought by the early saints to Manti was a coffee grinder. It was used to grind their grains into flour and meal because there was no mill available.

Manti Temple Whole Wheat Bread

12 C. wheat flour
4 C. white flour
8 C. warm water
1 C. melted shortening
1 C. powdered milk
1 1/2 T. salt
6 T. honey
2 pkgs. yeast

Mix together and knead 10 min. Let rise twice and put in pans. Let rise in pans no longer than 15 min. Bake at 350 degrees for 50 min. Yield 6-7 loaves

Vicki Tate

100% Whole Wheat Bread

1 1/2 C. warm water
3 T. honey
1 can evaporated milk
1 T. salt
1/3 C oil
2 eggs beaten
2 T. yeast
7 C. flour

Mix in order and raise in greased bowl, 35-45 min. Punch down, divide into 3 loaves. Raise for 20 minutes. Bake 350 degrees -45 min. This dough can be used for dinner rolls and cinnamon rolls.

Louise P. Eddy

Old Time Whey Bread

Mix:
3 C. whey milk (warmed)
1 1/2 T. yeast
2 T. honey
Add:
2 C. flour and mix
Add:
2 eggs
1 cube melted margarine or 1/2 C. shortening
3/4 t. salt

Mix well. Add flour till desired consistency (about 4 C.) Let rise. Form into loaves, rolls or bread sticks. Bake at 375 degrees.

Debra Lindsay

Pinto Bean Bread

2 pkgs. yeast
1/2 C. lukewarm water
1/2 C. evaporated milk
1 T. salt
6-7 C. flour
1 1/4 C. luke warm bean juice
1 C. mashed pinto beans
2 T. shortening
2 T. sugar

Soften yeast in lukewarm water. Heat bean juice, add milk, oil, salt, sugar, mashed beans and 2-3 C. flour. Beat until smooth. Add yeast and more flour until dough is easily handled. Knead until smooth and satiny. Place in greased bowl and let rise till double in bulk. Punch down and let rise again. Shape and place in 2 greased bread pans. Let rise again. Bake 40-45 min at 375 degrees.

Mary Fehlberg

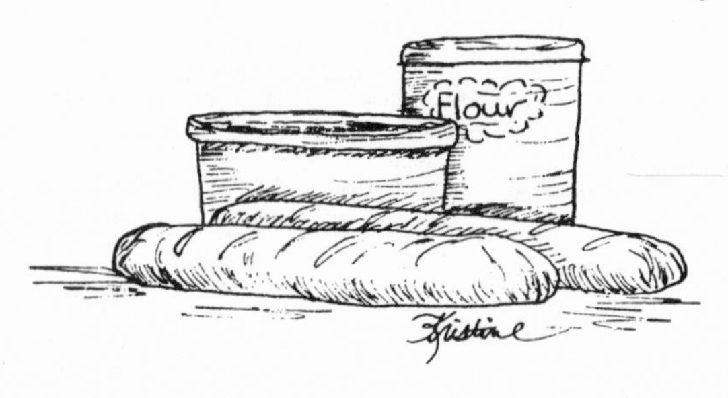

Oatmeal Bread

(Makes 2 Loaves)

1/2 C. warm water
2 T. active dry yeast
3/4 C. boiling water
3/4 C. rolled oats
1 C. buttermilk
1/3 C. oil
1/2 C. honey
2 C. flour
1 T. salt
1/2 t. soda
3 - 3 1/2 C. flour

Stir yeast into 1/2 C. water and allow to stand and bubble up. In sauce pan mix 3/4 C boiling water with oatmeal and cook a few min. Remove from heat and add buttermilk, oil, and honey. Sift flour, salt and soda in bowl. Add yeast and oats. Beat well. Gradually add flour to make a stiff dough. Flour a cutting board and knead for 10 min. Let dough rise 1 1/2 hours punch down and form into loaves and put in bread pans and let rise till double. Bake at 375 degrees for 45 - 50 min.

Grandma's Wonderful White Bread

(Old Recipe)

flour 2-3 sifters
2 pkgs yeast
4 T. sugar
4 T. shortening
2 T. salt
5 C. scalded milk and cooled to lukewarm

Add yeast to scalded and cooled milk. Add other ingredients and mix till an elastic ball. Knead for 10 min. Put in well greased bowl and let rise till triple in size, about 2 hours. Knead down and let set for 20 min. Form loaves. Let rise till light. Bake at 350 degrees for 15 min. then 300 degrees for 30-40 min.

This is my Grandmother Cowan's recipe and has been treasured by my family for many years.

It wasn't until the Danish people immigrated to Utah that yeast was brought to raise the breads. Before then they made a starter which they called "Sourdough". It was made by combining flour, a little salt, and enough warm water to make a spongy dough. This was put in a crock with a loose lid and kept warm for several days, during which time it bubbled and formed its own yeast. These breads weren't as light as yeast breads but the wonderful flavor made up for it. Sourdough was a favorite of the sheepherders in Sanpete County and still is.

Sourdough Start

Start:
2 1/2 C. milk
2 1/2 C. flour
1/2 yeast cake
1 t. sugar

Let set two days in a warm place to work. Save 1 cup and put in pint jar with a tight lid. Refrigerate.

Anna Jean Hedelius

Changing Any Recipe To Sourdough

Almost any recipe can be changed to use sourdough. To make it work you have to control the leavening and keep the thickness or moisture the same. For cookies, cakes and pastries add 1 C start and 1 t. soda. If the recipe calls for baking powder, then leave a teaspoon of it out. If the recipe doesn't call for baking powder but the results are too heavy, put a little baking powder in it.

When you use the sourdough start it increases the liquid and flour, so take out 1/2 C. of flour from the recipe to compensate. Buttermilk can be substituted for the mlk. Buttermilk and sourdough compliment each other.

Dehydrating Your Sourdough Start

Your sourdough start can be dried. Spread a thin layer on a piece of plastic wrap. Dry it in a food drier or just air dry it. Peel it off and turn it over to let it dry on the other side. When it is dry and brittle, break it into pieces and grind or blend in a blender. Store it in the refrigerator in a plastic bag or airtight container. To get your start going again just add water or milk to regain the original consistency. When using the reactivated start, let it sit at room temperature for 8 hours or so . By drying your start you can take it on a backpacking trip . Just mix the dry start with the flour for biscuits or pancakes. The night before you want to use it add the milk or water and let it set overnight. Then add the rest of the ingredients as you are making it in the morning.

Sourdough Starter
(With Yeast)

Starter
1 T. active dry yeast
2 1/2 C. flour
2 1/2 C warm water

Dissolve yeast in warm water. Add flour. Mix well. Loosely cover with a lid and let stand in a warm place for 3 to 5 days

Feeder
Add 2 1/2 C. flour
2 C. warm water

Allow to stand overnight at room temperature. Always save 1 cup to keep the starter going.

Sourdough Starter
(without yeast)

2 C. luke warm potatoe water
2 C. flour
1 T. sugar or honey

Make potatoe water by cutting up 2 potatoes and boiling in 3 cups of water until tender. Remove the potatoes and measure 2 cups of liquid. Mix water, flour and sugar into a smooth pasty sponge. Set in a warm place for several days. It should double its original size.

Sourdough Pancakes

4 C. sourdough Starter (2 T sugar)
1 egg
2 T. melted butter
1/4 C. evaporated milk or cream
1 t. salt
1 t. soda

Mix starter, egg, butter and milk. When it is well beaten, add remaining ingredients. Beat. Thicken with flour if needed. Fry on a griddle.

Sourdough Biscuits
(Pioneer Recipe)

1 C. starter
1 t. soda
1 t. salt
1 t. sugar
1 T. shortening
3 1/2 C. sifted flour

Place the flour in a bowl, make a hole in the center and add starter. Stir in all other ingredients. Gradually mix more flour to make a stiff dough. Pinch off enough dough to form a ball. Roll it in melted butter or shortening and arrange in a cake pan. Let rise 20 min. Bake at 425 degrees until done approximatly 15 min.

Sourdough Bread
(With Yeast)

1 C. starter
1 pkg. yeast
1 1/2 C. warm water
1/2 t. baking soda
6 C. unsifted all-purpose flour
2 t. salt
2 t. sugar

Sprinkle yeast over warm water. Stir in the sourdough starter, 4 C. of the flour, salt, and sugar. Stir vigorously for 3 min. Transfer to a large greased bowl, cover with a tea towel, and let rise in a warm place until double in size. This takes about 2 hours. Mix baking soda with 1 C. of the remaining flour and stir into dough. Turn dough onto a floured board and knead in remaining cup of flour (more if necessary) until dough of smooth and not sticky. Shape it into one large round loaf or 2 oblong loaves and place on a lightly greased cookie sheet. Cover and let rise in a warm place until nearly double in size. Before baking, brush surface with water and score or slash the top diagonally with a sharp knife. Before putting bread in the oven to bake, place a shallow pan of hot water in the botton of the oven. Bake bread in a preheated 400 degree oven for 45 to 50 minutes.

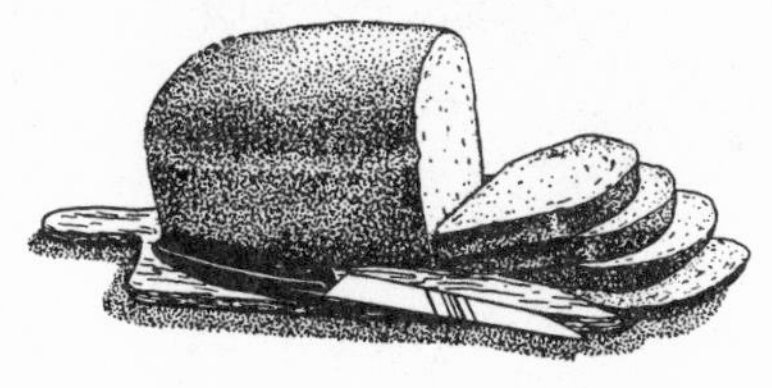

Sourdough French Bread

1 1/2 C. warm water
1 pkg. dry yeast
1 C. sourdough start
4 C. unsifted flour
2 t. sugar
2 t. salt
1/2 t. soda
About 2 C. unsifted flour

Pour warm water into a large mixing bowl and stir in the yeast. Add the start, 4 C. of flour and sugar. Stir vigorously for 3 or 4 minutes. Cover and place in a warm spot until it nearly doubles in bulk (about 2 hours). Mix soda and salt with one cup of flour and stir in. Turn the dough out into 1 C. of flour spread on a bread board. Coat with flour and knead until satiny adding more flour if necessary

Shape into two oblong loaves or one large round loaf. Place on a lightly greased cookie sheet, cover and put in a warm place to raise. When nearly doubled in size brush lightly with salad oil and make diagonal slashes across the top with a single edge razor blade. It takes about 2 hours for the dough to double in size and it is delicate, so brush the oil gently and make the slashes clean and shallow or it will fall.

Bake in a preheated oven at 450 degrees until the crust is a medium dark brown. Oblong loaves about 35 to 40 minutes. For a tougher crust, take the bread from the oven 10 minutes early and wipe the crust with salt water then return to the oven to finish baking.

Sandy Clayton

Sourdough Pizza Crust

1 C. sourdough start
1 t. salt
1 T. melted butter
1 C. florr (about)

Mix together start, salt and butter. Work in flour so that dough can be rolled into a thin layer. Add a little more than a cup of flour if necessary. Roll out on a floured board or pat to fit an oiled pizza pan. Bake at 500 degrees for 10 minutes. Spread the baked dough with your favorite pizza sauce and trimmings. Bake 425 degrees until sauce bubbles, about 15 minutes.

Sandy Clayton

Sourdough Buttermilk Pancakes

This excellent recipe makes the lightest panckes we have ever tried and no sponge is needed the night before.

1 1/2 C. starter
2 eggs
2 T. sugar
2 T melted butter or oil
1 1/2 t. salt
1 t. soda
1 C. buttermilk
1 C. flour

Put the start in a warm bowl and let it set till it is room temperature. Separate the eggs and stir in the yolks with the start. Warm the buttermilk to lukewarm and stir it and the butter and flour into the start. Combine the sugar, salt and soda and sprinkle it over the top of the batter. Fold it in with a large spoon. Beat the egg whites until they peak and fold them into the batter. Nothing left to do but cook and enjoy.

Sandy Clayton

Sourdough English Muffins

1 C. starter
2 1/2 C. flour
3/4 C. buttermilk
1/4 C. yellow cornmeal
1 t. soda
1/2 t. salt

Take buttermilk and start from the refrigerator at least 30 minutes before preparing recipe. When all ingredients are at room temperature, combine and stir well. Turn dough onto lightly floured board and knead until smooth, adding more flour if necessary. Roll dough into 1/2 inch thickness and allow to rest for 10 minutes. Cut with a floured 3 inch biscuit cutter. A tuna fish can with the ends removed works well. Sprinkle 2 Tablespoons of cornmeal on a sheet of waxed paper. Place muffins on cornmeal and press lightly to coat each side. Cover with plastic wrap and let rise for 1 hour. Fry on a lightly greased griddle heated to 350 degrees for 20 minutes. Turn them occasionally while cooking.

Sandy Clayton

Sourdough Chocolate Cake

2/3 C. shortening
1 2/3 C. sugar
3 eggs
1 C. start
2 C. flour
2/3 C. cocoa
1/2 t. baking powder
1 t. salt
3/4 C. water
1 t. vanilla
1 C. chopped nuts (opt.)
1 1/2 t. soda

Cream shortening and sugar. Add eggs, one at a time beating after each addition. Stir in start. Mix together flour, cocoa, baking powder, soda, and salt. Add alternately with water and vanilla. Mix at low speed. Stir in walnuts. Pour into two 9 inch cake pans that have been greased and lightly floured. Bake at 375 degrees for about 35 minutes or until cake tests done. Allow to cool for 10 minutes, then invert on cooling rack, removing pans carefully. Cool thoroughly and frost with a favorite chocolate frosting.

Sandy Clayton

Sourdough Trail Bread

When you get up in the morning with a hankerin' for hot bread and don't have anything started, here's a way to make it quick and easy.

Use 1 C. of sourdough start, 1 t. of baking powder, 1 t. of soda, 1/4 t of salt and enough flour to make a stiff dough. Stir or knead them together until they are mixed well. Roll the dough out or flatten it out real thin, less than a quarter of an inch. Cut pieces small enough to fit in your frying pan. Use plenty of bacon grease, butter or whatever you have to cook it in. Cook over a slow fire until golden brown on the bottom. Turn it over and do the same on the other side. Serve it hot.

If you happen to be caught without a pan to fry it in, just wrap the dough around a stick and bake it over hot coals. Try it on your stove.

Sandy Clayton

Corn was an important staple to the early Sanpete pioneers. Sometimes they used it whole as in corn chowder, but more often it was ground into cornmeal and made into corn bread, corn tortillas, corn muffins, and mush.

Mom's Favorite Corn Bread

1 C. cornmeal
1 C. flour
2 T. sugar
1 T. baking powder
1 t. salt
1/3 C. oil
1 egg
1 C. milk

Measure milk and put in a bowl. Add to milk: egg, oil, and sugar. Beat. Sift flour, baking powder, salt. Stir into milk mixture and add cornmeal. Mix well.

Pour into baking pan and bake at 400 degrees for 25 min.

It's excellent to serve with beans. The two combined make a complete protein. My children especially like it with hot chocolate for breakfast.

Elaine Westmoreland

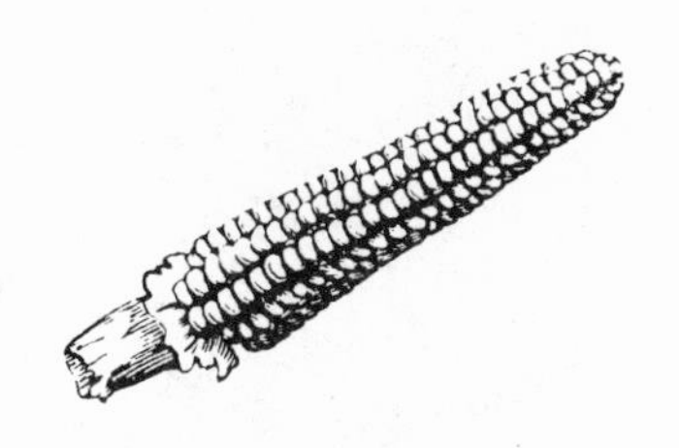

Sweet Corn Bread

1/2 C. butter
1 C white flour
1/2 C. cornmeal
1/4 C. sugar
2 t. baking powder
Pinch of salt
1-2 eggs
1/4 C. milk

Melt butter. Sift dry ingredients together into a bowl. Beat egg and milk together. Add butter and stir into flour mixture. Pour batter into baking pan or muffin tins. Bake 15-20 min. at 350 degrees.

Variation: Add 3/4 C. diced apples or apricots.

Stephanie and Kimberly Tate

Cornmeal Fritters

2 C. milk
1/4 C sugar
4 eggs
1/2 C. corn meal
1/2 t. salt
nutmeg (if desired)

Cook milk and meal together 15 min with salt and sugar. When cool add the eggs well beaten. Drop into hot fat and fry.

Mary Fehlberg

Corn Bread from Mix

Master Mix on page 75

1 1/3 C. milk
4 1/2 C. Cornmeal Mix
2 eggs, beaten

Stir till moistened. Bake at 425° for 25 min.

Mary Fehlberg

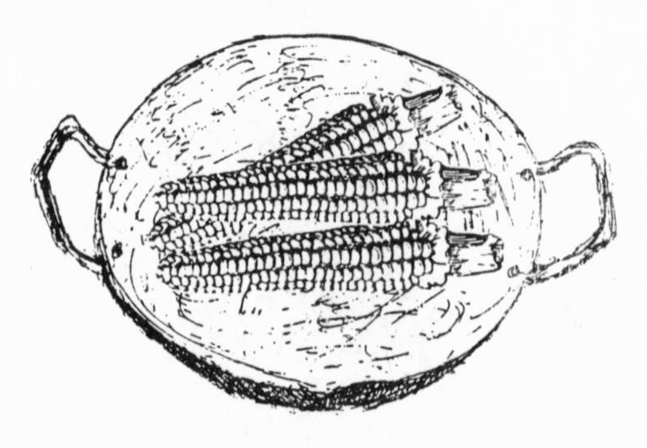

Ash Cake

(Pioneer Recipe)

4 C. corn meal
1 T. lard (shortening)
2 t. salt
boiling water

Scald meal in water. Add the salt and shortening. When mixture is cool form into oblong cakes, adding more water if necessary. Wrap the cakes in cabbage leaves, or place one cabbage leaf under cakes and one over them, and cover them with hot ashes.

The Indians cooked Ash Bread by making a hole in the center of the ashes of a hot fire, raked down to the hearth, then placing the dough in the hole, let it crust and cover with hot ashes and embers. Bake to suit taste

Hoecake

(Pioneer Recipe)

Hoecake---similar to Corn Pone and Ash Cake--was baked on a hoe over an open fire.

1 C. water-ground white corn meal
1/2 t. salt
1 T. lard, melted
boiling water

Combine corn meal and salt, then add lard and enough boiling water to make a dough heavy enough to hold a shape. Form into 2 thin oblong cakes and place in a heavy, hot well-greased pan. Bake in a preheated 375 degree oven about 25 minutes . Serve hot. May be cooked slowly on both sides on a well greased griddle.

Corn Pone

(Indian Recipe)

2 C. white corn meal
1 t. salt
1/4 t. baking soda
4 T. shortening or lard
3/4 C. boiling water
1/2 C. buttermilk (about)

Sift together corn meal, salt, and baking soda. Work in fat with finger tips until well blended. Pour in boiling water and continue to work the mixture. Gradually add enough butter milk to make a soft dough, but one firm enough to be molded or patted into small, flat cakes. Place cakes in a hot well-greased iron skillet and bake in a preheated 350 degree oven for 35 minutes

Grandma's Refrigerator Rolls

(Old Recipe)

1 C. hot water
1 t. salt
1/4 C. sugar
1 pkg. yeast
1 egg
6 T. shortening
3 1/2 - 4 C. flour
2 T. luke warm water

Soak yeast in water. Mix together hot water, salt, shortening, sugar and egg. Add yeast mixture. Sift in enough flour to make easy to handle. Knead on floured board. Place in bowl and let rise till double. Roll out and shape into rolls. Let rise. Bake in hot oven for approx. 20 min.

Grandma would always make these for every family dinner. Our get-togethers just weren't the same without grandma's famous rolls.

Old Fashioned Sticky Buns

Master Mix on page 73

Wheat Quick Mix biscuits
margarine
cinnamon
sugar

Topping
1/2 C. brown sugar
1/4 C. canned milk
2 T. butter
Nuts
Raisins
2 T corn syrup

Prepare biscuit recipe. Roll dough into rectangle and spread with softened butter. Sprinkle with cinnamon and sugar and roll as for jelly roll. Cut into 10 slices.

In a sauce pan, bring to boil, brown sugar, canned milk, corn syrup. Pour into buttered baking dish. Sprinkle with nuts and raisins. Lay biscuit slices on top. Bake 375 degrees 15-20 min.

Elaine Harmston

Hamburger Buns

Mix and let rest 10 minutes
3 1/2 C. warm water
1 C. oil
3/4 C. sugar or 1/2 C. honey
6 T yeast
Add:
1 T. salt
3 eggs, beaten
10 - 12 C. flour or wheat flour

Knead for 5-10 minutes . Shape into buns. Place on cookie sheet and brush tops with butter. Let rise for 10-15 minutes. Bake at 425 degrees 12-15 minutes.

Elaine Harmston

Buttermilk Rolls

1 C. buttermilk
1/4 t soda
3 T. soft shortening
2 1/2 C. flour
1/4 C. sugar
1 t. salt
1 T. yeast in 1/4 C. warm water
1 egg

Heat buttermilk to lukewarm. Add shortening, yeast, sugar and slightly beaten egg. Add salt and soda to flour, mix well. Add to buttermilk mixture. Let rise until double in bulk. Make into rolls and let rise again. Bake at 375 degrees until slightly brown. Makes about 2 dozen rolls.

Nora Mickelson

Cinnamon Rolls from a Mix

Master Mix on page 73

2 C. "Biscuit" Mix
2 t. cinnamon
4 T. margarine (melted)
raisins
1/3 - 1/2 C. water
3 T sugar
1 egg, beaten

Combine mix, egg, water. Mix gently with fork. Turn onto floured surface. Roll to 1/4" thickness. Spread with margarine and sprinkle top with sugar and cinnamon and raisins. Roll as for jelly roll. Cut into 1" slices. Place on greased pan. Bake in 425 degree oven for 15-20 min.

Lisa Conn

Quick and Easy Buttermilk Scones

2 C. buttermilk
4 t. active dry yeast
5-6 cups all purpose flour
2 t. baking powder
1/4 C sugar
1/2 t. soda
1/2 C warm water
1 t. salt
2 eggs
2 T. vegetable oil for frying

Warm the buttermilk. Dissolve the yeast in 1/2 C. warm water. Sift together the flour, salt, soda, and baking powder. Beat the eggs in a separate bowl. Add the sugar and oil to the eggs. Next add the buttermilk and yeast mixtures. Gradually add sifted, dry ingredients. Beat constantly. Use a wooden spoon to stir with. Add enough flour to make a soft dough. Turn onto a floured counter and roll out the dough to about 1/2 ' thick. Cut it into squares. Let it rise for a half hour. Heat the frying oil , and fry the scones to a golden brown.

Best Ever Scones

2 C. hot water
1 T. oil
2 1/2 T. egg mix*
7 T. powdered buttermilk
1 1/4 t. baking powder
1 T. sugar
6 C. flour
1 T. yeast
3/4 t. salt
1/4 t. baking soda

In large bowl mix water, sugar, and oil. Mix dry ingredients together. Add liquid to dry ingredients and knead into a soft dough. Let rise. Divide into 2 sections. Pat out on floured board . Cut into pieces about 4"x4" deep fry.

*If using 1 fresh egg, beat it well and add to hot water with about 1 cup of the dry ingredients.

Cathi Call

Indian Frybread

(Old Indian Recipe)

2 C. bleached enriched flour
1/2 C. powdered milk
1 1/2 tsp. salt
1 T. baking powder
3/4 C. warm water
2 T. shortening

Mix dry ingredients and shortening together. Rub mixture with fingers until coarse crumbs form. Push to one side of a big bowl and add water. Mix in a circular motion with your fingers, then knead in an upward motion, stretching it upward with one hand. This should make a soft dough.

Frybread: Knead on a floured board for 2-3 minutes. Divide into 6 portions and shape into balls. Flatten dough with fingers from the center outward until it's 6 - 7 inches around. Pat it back and forth with hands like you make tortillas. Thin out edges so it's all the same thickness. Fry in hot oil until edges brown. Turn over and continue frying until brown. Fry bread can be cooked outdoors, too, over an open fire.

Tortillas: Use the same recipe and the same technique as fry-bread, but make them thin. Fry on medium hot griddle with oil, turning to brown both sides.

Bread that's Cooked Over Hot Coals: Use the same recipe and method as for frybread. Pat out to medium thickness and cook over a wire rack that's been placed over hot coals. Turn to brown both sides.

Note These breads are delicious served with chili, cheese, lettuce, tomatoes, salsa, and sour cream. Also delicious eaten with butter, honey, or jam. May also be served with stew or other foods as a plain bread.

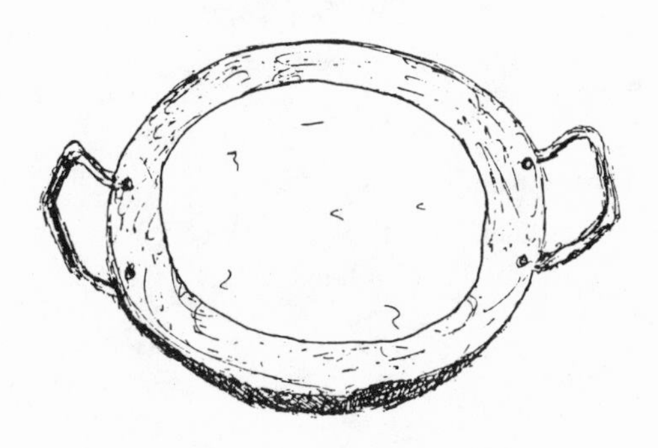

Indian Fried Scones

(Very easy fry bread)
(Early Settlers)

3 C. flour
4 t. sugar
4 t. baking powder
2 t. salt
1 1/4 to 1 1/2 C warm water

Mix together and let rise about 5 minutes. The least it is handled the better. Shape as scones and fry to golden brown. Serve with honey butter or jam.

Elta Alder and Ruth Scow

Indian Bean Bread

(Early Settlers)

4 C. corn meal
2 C. hot water
2 C. cooked beans
1/2 t soda

Put cornmeal in a bowl, mix in drained beans. Make a hole in middle and add soda and water. Mix. Form into balls and drop into a pot of boiling water. Cook about 45 min or till done.

Navajo Bread

4 1/2 C. flour
1 T. soda
1 1/4 C. warm water
2 T. powdered milk
1 t. salt

Mix and let stand 30 minutes. Break off pieces and roll into balls about 2" in diameter. Pat or roll the balls flat into tortilla shape. Deep fry until crisp.

Lisa Conn

Brigham's Buttermilk Doughnuts

(pioneer recipe)

2 C buttermilk
2 eggs, beaten
5 C flour
1 C sugar
2 t. soda
1 t. baking powder
2 t. salt
1 t. nutmeg
1/4 C melted butter or oil

This recipe came from Brigham Youngs wife, Emily Dow Partridge Young.

Combine the buttermilk, eggs and sugar. Mix well. Add sifted dry ingredients and melted butter. Roll to 1/4 " thick. Cut with a doughnut cutter. Fry in hot oil until brown. Drain and sprinkle with sugar.

Elta Alder

Unleavened Bread Sticks

(Early Settlers)

4 C. wheat flour
1 T. salt
1/2 C. oil
1. C milk
3 T. brown sugar or honey

Mix together flour, salt, oil. Add milk and sugar or honey.
Knead a little and roll into sticks the size of your finger. Bake on cookie sheet at 375 degrees about 20 min.

Mary Fehlberg

Wheat Bagels

4 C. wheat flour
1 T. yeast
2 T. oil
1 T. brown sugar
1 qt. boiling water
1 1/ 2 t. salt
5 1/3 T. honey
2 eggs, beaten
1 C. potato water(water poured off boiled potatoes)

Sift flour and salt into bowl. Soften yeast in 1/3 C. warm potato water and stir into flour. Add the honey and oil to remaining potato water and stir into flour mixture. Add eggs. Knead for 10 minutes adding flour if necessary.

Put dough in oiled bowl and let rise till double (1 1/2 - 2 hours). Preheat oven to 450 degrees. Knead dough again till smooth and elastic. Pinch off dough and roll till 6" long and 3/4 " thick. Bring ends together and pinch to form doughnut shape.

Drop sugar into boiling water. Drop bagels into water one at a time. When they come to the top, turn them over and boil 1 min longer.

Bake on cookie sheet 10-15 minutes till golden brown.

Vicki Tate

Banana Bread

1 C. oil
1 C. honey
4 eggs
2 t. soda
1 t. salt
5 large bananas
4 c. flour

Mix in order as given. Pour into 2 bread tins. Bake 350 degrees 45-40 minutes.

Louise P Eddy

Old Fashioned Pear Bread

1 can pears
1/2 C. sugar
1 t. salt
1/4 C. oil
1 C. powdered sugar
1/2 C. chopped walnuts (opt)
2 1/2 C. flour
3 t. baking powder
1/8 t. nutmeg
1 egg, beaten
1-2 T. Tang or orange drink

Drain pears, reserving syrup. Reserve 1 pear half for topping. Mash pears till smooth. Transfer to 1 cup measure. Add as much pear syrup as needed to make 1 cup.

Mix dry ingredients in large bowl. Mix pear puree, oil, egg in medium bowl. Add to flour mixture. Stir till just moistened. Add nuts if available. Batter will be stiff. Pour into greased loaf pan. Cut reserved pear half into 6 slices. Arrange crosswise on row on top of batter. Bake 50-55 minutes at 350 degrees. Mix powdered sugar and enough orange drink to make thin glaze. Frost. Let stand overnight.

Vicki Tate

Banana Bread
(Dried food)

2 C. flour
1 t. soda
1/4 C. water
1/2 C. oil
1/2 C. dehyd. egg mix
1 C. chopped nuts (opt)
1 C. crushed dried banana
1/4 t. salt
1/2 t. vanilla
1 C. sugar
1/3 C. water

Mix egg mix in 1/3 C. water. Cream oil and sugar. Add egg mixture and blend. Sift dry ingredients. Mix banana, water and vanilla. Add to creamed mixture alternating with dry ingredients. Blend well. Pour into greased bread pan. Bake at 350 degrees for 1 hour.

Natalie Simmons

Zucchini Bread

3 eggs
2 C. zucchini*
1 1/2 C. honey
2 t. baking powder
3 t. cinnamon
1 C. chopped walnuts
2 t. vanilla
1 C. oil*
1 t. baking soda
4 1/2 C. wheat flour
1 t. salt

*Note: I cut the zucchini in cubes and put in blender with oil and blend until smooth.

Beat eggs until light. Add next four ingredients and mix well using electric mixer. In a seperate bowl mix 4 1/2 C. whole wheat flour, cinnamon and salt. Blend the dry ingredients with the egg mixture and 1 C. chopped walnuts. For variety add any combination of the following ingredients.

1 C. raisins
1 C. dried cherries
1 can crushed pineapple (increase flour by 1/2 C.)

Bake 325 degrees for 1 hour approx. Makes 2 loaves.
This freezes very well!

Debbie Wade

Cheese Puffs

4 T. margarine
3/4 C. flour
1/2 C. cheddar cheese
dash of cayenne pepper
3/4 C. water
2 eggs
1 t. salt

Bring butter and water to a boil. Mix in flour and stir until dough is a solid ball. Remove from stove and beat in 1 egg. Then add another egg and beat until glossy and well blended. Beat in cheese, salt, and cayenne. Drop by spoonfuls on cookie sheet. Bake at 375 degrees about 25-30 minutes. Will be puffy and crisp.

Pita Bread

Dissolve:
1 pkg. yeast
2 1/2 C. warm water
Add:
2 t. salt
5 1/2 C. wheat flour

Mix and knead by hand 10 minutes on floured surface. Let rise till double in size. Divide into balls size of walnuts. Roll into circles 1/4" thick. Let stand 10 minutes. Bake on cookie sheet. 6-8 minutes at 450 degrees.

Jill Hanson

Breakfast Sweet Bread

Master Mix on page 73

2 C. Wheat Quick Biscuit Mix
2 eggs beaten
1/2 C. raisins
1/2 C. chopped nuts
1/4 C. brown sugar
3/4 C. water
1/2 brown sugar

Blend mix, sugar, eggs, and milk. Add raisins. Pour into greased 9" square pan. Combine brown sugar, cinnamon and nuts and sprinkle on top. Bake 350 degrees 35-40 min. Serve hot.

Elaine Harmston

Mormon Muffins

2 C. boiling water
2 C. raisins
4 t. soda
1 qt. buttermilk
5 C. flour
1 T. salt
1 C. shortening
1 1/2 C. sugar
4 eggs
4 C. All Bran flakes or Raisin Bran

Pour boiling water over raisins and cool 5 min. Then add 4 t. soda and set aside.

Cream together shortening, sugar, eggs, then add buttermilk and remaining ingredients. Mix this mixture with the raisin, water and soda mixture. Store in a covered container. So not stir. Bake in muffin tins at 350 degrees for 10-15 min. Can store this in refrigerator for up to 4 weeks.

Peggy Layton

Country Hearth Apple Muffins

2 C. flour
2 t. baking powder
2/3 t. cinnamon
1/2 t. salt
1/2 t. nutmeg
1/2 C. sugar
1 egg
1 C. milk
4 T. butter
1/2 C. nuts
1/2 C. finely chopped apple

Sift together dry ingredients. Combine melted butter, egg, milk and stir well into dry mixture. Fold in apples and nuts. Fill muffin tins 2/3 full.

Topping: 1 t. cinnamon, 1/2 C. sugar, combine, sprinkle generously over each muffin. Bake 30 min at 400 degrees.
Delicious!

Vicki Tate

Healthy Oatmeal Muffins

1 C. buttermilk
1 egg
1 C. whole wheat flour
1/2 t. soda
1 C. rolled oats
1/2 t. salt
1 1/2 t. baking powder.
1/4 C. vegetable oil
1/2 C. packed brown sugar or 1/2 C. applesauce or frozen orange juice.

Pour buttermilk over oats. Let stand 5 minutes Add egg and brown sugar to oats and mix well. Add sifted dry ingredients and oil.. Mix well. If raisins or dates or other fruit such as chunks of apple etc.... are desired add them.

Spoon into 12 greased muffin tins or cup cake liners Bake at 400 degrees for 18 min. or until brown. Makes 12.

Peggy Layton

Queen of Muffins

Cream:
1/4 C. butter or shortening
1/3 C. sugar
1 egg
Add to creamed mixture and mix well.
2 C. flour
2 1/2 t. baking powder
1/2 t salt
1/2 C. milk

Bake 15 minutes at 400 degrees.
Bake in Microwave 4-6 minutes--turn every 1 1/2 -2 minutes

Anna Jean Hedelius

Granola Muffins

1 3/4 C flour
1 T. baking powder
1/4 t. salt
1/2 C. brown sugar
1 egg
1 C. milk
1/4 C. raisins
1 C. granola
1/2 C. whole wheat flour
1/4 C. cooking oil

Mix dry ingredients together. Combine egg, oil, and milk. Add to dry ingredients, stirring just enough to moisten. Spoon batter into greased muffin tins. Bake in 400 degree oven 20 min. About 12 muffins

Connie Gardner

Grandpa's Favorite Bran English Muffins

2 C. wheat flour
3/4 t. salt
1 C. milk
3 T. oil
2 C. white flour(approx.)
1 T yeast
1 T brown sugar
1/2 C. water
1 C. bran

Mix 1 C. wheat flour, yeast, salt and sugar in large bowl. Scald the milk, water and oil. Stir into first mixture. Beat with mixer for 2 minutes.

Stir in remaining wheat flour, bran, and enough flour to make dough firm. Knead for 10 minutes or until elastic. Put in oiled bowl and cover with plastic wrap. Let rise till doubled. Turn dough out, roll to 3 /4 " thickness. Cut with 3 " cutter. Let rise 25-30 minutes.

Put on lightly oiled hot griddle and bake 10-12 minutes on each side until lightly browned. Cool. When ready to serve, split, toast and butter.

Vicki Tate

Bran Muffins with Raisins

1 C. bran
1 egg, beaten
2 T. oil
1 1/4 C. wheat flour
1/2 C. raisins
1 C. buttermilk
1/4 C. honey
1 t. vanilla
1/2 t. salt

Soak bran in buttermilk for 5 -10 min. Stir in beaten egg, honey, oil, and vanilla. Mix well. Sift together flour, baking powder, and salt. Gently stir flour mixture and raisins into bran mixture. Batter will be slightly lumpy. Pour into muffin tins. Bake at 375 degrees for 20 - 25 minutes. EXCELLENT!

Vickie Tate

Apple Oatmeal Muffins

1 1/4 C. wheat flour
1 C. oatmeal
2 t. baking powder
3 t. cinnamon
1 t. soda
2 eggs
2 T. honey
1/2 C. milk
2 C. diced apples or raisins
1 1/2 t. nutmeg
1/2 t. salt
1/4 C. oil

Sift dry ingredients together. Combine wet ingredients and stir into dry ingredients. Mix well. Stir in apples or raisins. Fill muffin tins. Bake 15 - 20 minutes at 350 degrees.

Susan Westmoreland

Sugared Muffins from Mix

Master Mix on page 75

2 C. cornmeal mix
1 egg, beaten
1/2 C. milk

Combine ingredients. Stir to moisten. Bake in muffin tins 15-20 min at 425 degrees.

Dip top of hot muffins in 1/4 C. melted butter, then in 1/4 C. sugar and 1/2 t. cinnamon mixture.

Mary Fehlberg

Buttermilk Biscuits

2 C. flour
2 t. baking powder
1/3 C. buttermilk powder
1 t. soda
1/3 C. oil
2/3 C. water

Reconstitute buttermilk powder in water. Add oil. Sift together dry ingredients. Combine dry and wet ingredients. Stir with fork till moistened. Knead 10-12 times. Roll out to 1/2 " thickness and cut biscuits. Bake 450 degrees for 10-12 min.

Vicki Tate

Baking Powder Biscuits

2 C. flour
1/2 t. salt
2/3 to 3/4 C. milk
1 T. baking powder
1/4 C. shortening

Cut shortening into sifted dry ingredients till like coarse crumbs. Make a well in middle and add milk. Stir quickly with fork, only until dough follows fork around bowl.

Turn dough onto lightly floured surface. Knead 10-12 times. Roll out dough to 1/2 " thick. Cut dough. Bake on baking sheet at 450 degrees for 12-15 min.

Vicki Tate

Sky High Biscuits

2 C all-purpose flour
1 C whole wheat flour
4 1/2 t. baking powder
2 T. sugar
1/2 t. salt
3/4 t. cream of tartar
3/4 C butter or margarine
1 egg beaten
1 C milk

In a bowl combine the first 6 ingredients. Cut in the butter until it resembles cornmeal. Add the eggs and milk, stirring quickly and briefly. Knead lightly on a floured board. Roll or pat gently to a 1-inch thickness. Cut into 1 to 2-inch biscuits. Place in a greased 9-inch square pan. Bake at 450°F for 12 to 15 minutes.

Angel Bisquits

5 C. flour
1/4 C. sugar
2 tsp baking powder
1 t. salt
1 t. soda
1 C. shortening
2 C. buttermilk
1 T. yeast
2 T. water

Mix yeast and water together, set aside. Mix all dry ingredients together, then cut in shortening. Add buttermilk and yeast mixture,mix lightly. Roll to 1 - inch thick, cut into desired shape and place on baking pan. Bake 425° for 15 - 20 minutes.

Biscuit Ideas

Herb Biscuits: Add 1 t. caraway seed, 1/2 t. sage and 1/4 t. dry mustard to the dry ingredients. You can mix up your own favorites, if you would like.
Chive & Yogurt Biscuits: Substitute 1 c. plain yogurt for the milk plus 1 T. and 1 t. chopped chives.
Cheese Biscuits:: Stir in 1/2 C. of your favorite cheese.
Cornmeal Biscuits:Use 1/2 C. cornmeal in place of 1/2 cup flour. Sprinkle cookie sheet with cornmeal; place biscuits on cookie sheet and sprinkle with more cornmeal.
Rye Biscuits: Substitute 3/4 C. rye flour for 3/4 C. white flour.

PREPARE AHEAD MASTER
Basic Biscuit Mix
(like Bisquick)

9 C. sifted flour
1/4 C. sugar
2 C. shortening
3 T. baking powder
1 T. salt

Mix until flour and shortening are thoroughly mixed together. Use in recipes calling for Bisquick.

LeAnne Beal

PREPARE AHEAD MASTER
Wheat Quick Mix Biscuit Recipe
(For when you're in a hurry)

4 C. wheat flour
1/3 C. baking powder
1/2 C. sugar
2 C. shortening
4 C. white flour
4 t. salt
1 1/4 C. powdered milk

Mix all ingredients and cut in shortening. Store in refrigerator.

Elaine Harmston

Quick Mix Biscuits

2 C. Wheat Quick Mix Biscuits
2/3 C. Milk

Mix, cut, and bake at 400 degrees for 10-15 min.

Elaine Harmston

PREPARE AHEAD MASTER
Whole Wheat Pancake Mix
(Dried Food)

24 C. whole wheat flour
8 T. baking powder
2 1/4 C. powdered eggs
4 T. salt
4 C. dry powdered milk

Mix ahead and store in cool dry place.

To use mix:

Stir until just mixed 2 C. dry pancake mix and 2 1/2 C. water (more for thinner batter) add 2 T. oil. Griddle on moderately hot, lightly greased griddle, turning when edges are set.

Basic Cookie Mix

6 C flour
2 C sugar
2 t. salt
1 C powdered Milk
2 T baking powder
1 1/2 C shortening

Sift all dry ingredients togerther. Cut in the shortening until it resembles coarse meal. Store in an airtight container.

Natalie Simmons

Oatmeal or Rolled Wheat
Basic Cookie Mix

4 C flour
1 C sugar
1 C brown sugar
2 t. soda
2 t. salt
2 C shortening
4 C oatmeal or rolled wheat
2 t. baking powder

Sift together all dry ingredients except oatmeal. Mix in the sugar. Cut in the shortening until mixture forms fine crumbs. Add oatmeal or rolled wheat. Mix well. Store in a tightly covered container.

PREPARE AHEAD MASTER
Corn Meal Mix
(For when you're in a hurry)

8 1/2 C flour
1/2 C. baking powder
3 C. powdered milk
3 C. shortening
***3 C. white cornmeal**
2 C. yellow cornmeal
2 T. salt

Sift dry ingredients together. Cut in shortening till well blended. Place in tightly closed container

*All yellow cornmeal can be used.

Mary Fehlberg

Carmen's Flour Tortillas

5 C. flour
1/2 C. shortening
1 t. salt
1 1/2 - 2 C. hot water

Mix all ingredients in a large bowl. Put out on floured board when a ball is formed. Knead until pliable. Pinch off pieces about 2 " in diameter and roll out to 6" - 8 " round tortilla. Cook in a dry cast iron skillet or on a griddle. Cook on each side until the bubbles that form are browned. Push down large bubbles when turning for the last time with a cloth. Makes about 20.

Connie Gardner

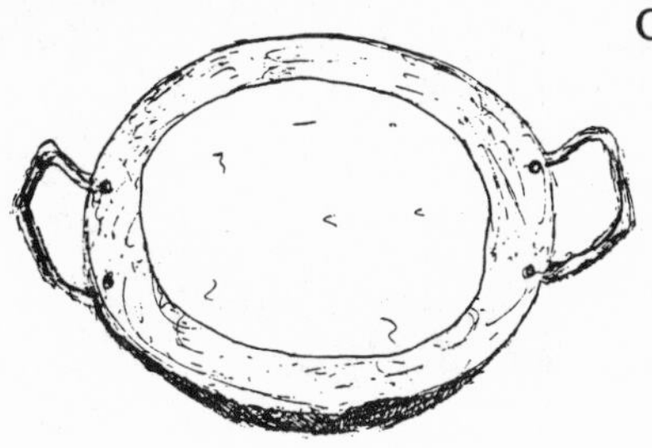

Whole Wheat Pizza Crust

1 C. wheat flour
1 t. salt
1 t. oregano
pinch pepper
2 eggs
2/3 C. milk
cornmeal

Beat eggs, add milk, then flour and seasonings. Dust pizza pan with cornmeal. Pour in batter, spreading to sides. Bake at 425 for 30 minutes.

Vicki Tate

Tortillas

1 1/2 C. cold water
1 c. flour
1/2 C. cornmeal
1/4 t. salt
1 egg

Beat all ingredients with a beater until smooth. Heat skillet over medium-low heat just until hot. Grease skillet with oil. Pour scant 1/4 C. of the batter into skillet; immediately rotate skillet until batter is about 6-inches in diameter. Cook about 2 minutes on each side.

Corn Tortillas

1 C. wheat flour or white
1 C. cornmeal
3/4 t. salt
2 T. oil
2/3 C. warm water

Knead several minutes. Divide dough into 12 balls. Roll the tortillas out thin. Heat 14" oil in skillet till very hot. Fry one at a time 35-40 sec.

Mary Fehlberg

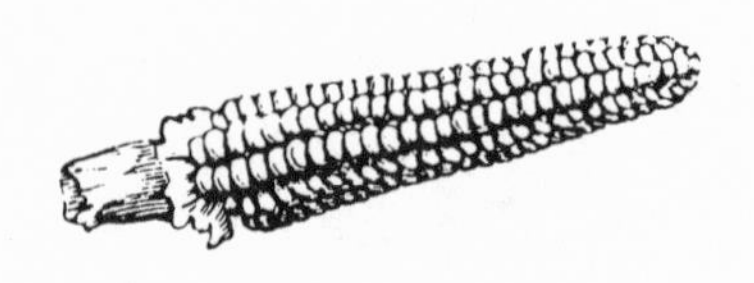

Wheat in a Thermos

1/2 C. wheat kernels
1 qt. boiling water

Place wheat in thermos and pour water to fill thermos. Screw on lid. Sit 2 hours or overnight. Strain.

Whole Wheat Cereal

2 C. water
1 C. whole wheat
1/2 t. salt

Cook on low overnight in crock pot.

Cornmeal Mush

1 C. boiling water
1 C. cold water
honey and milk
1 C. corn meal
1 t. salt

Pour the boiling water into a sauce pan. Mix together the corn-meal and cold water and gradually stir into the boiling water.

Add the salt and cook, stirring occasionally for about 20 minutes. Serve with honey and milk.

Vicki Tate

Pioneer Style Cold Cereal

2 C. wheat flour
1/2 t. salt
2 C. water
2 T. honey

Stir water, salt and honey into flour. Spread on cookie sheets (1/2 C. to each cookie sheet). Bake 15 min. at 350 degrees. Break into bite size pieces.

Cherie Harmon

Apple Cinnamon Crunch Cereal

Combine in large bowl:
4 C. rolled oats
1 C. nuts, chopped (optional)
3/4 t. salt
1/2 C. coconut
1/2 C. sesame seeds (optional
1 t. cinnamon

Mix seperately, then add to above mixture:
1/2 C. honey
1/2 t. vanilla.
1/3 C. oil

Mix thoroughly, spread on 2 large cookie sheets and bake at 350 degrees for 20 - 25 min., stir occassionally. Add 1/2 - 1 C. dried apples cut into fine pieces. Store in tightly covered container.

Vicki Tate

Almond Crunch Granola

10 C. rolled oats
4 C. sliced almonds
1 1/4 C. pure maple syrup
1 1/4 C. safflower oil
1 1/4 t salt
4-5 t. vanilla

Optional ingredients:
Any other nuts (2 C.)
raisins (2 C.)
dried cherries (2 C)
unsweetened coconut (2C)
Drained crushed pineapple (1 - 20oz can)
Any combination dried fruit

Mix all ingredients together. Spread thinly on cookie sheets. Bake 20-25 minutes at 325-350 degrees or until lightly browned. Cool 10 min. in pan then transfer to tupperware bowl. Keep tightly covered. This recipe makes a large (green or yellow Tupperware bowl) Will keep only about 2 months in cupboard because of the oil.. I Never have any left more than 3-4 weeks. Everyones favorite breakfast cereal and ice cream topping.

Debbie Wade

Peanut Butter-Oatmeal Granola

1/4 C. margarine
1/2 C. brown sugar
1/4 C. milk
2 C. oats
1 t. vanilla

1/4 C. peanut butter
3 T honey
1 T. cinnamon
1/2 C. 6 grain mix

In sauce pan on medium heat melt first 5 ingredients. Bring to boil. Let boil one minute, stirring often. Remove from heat. Add cinnamon , oats, 6-grain and vanilla. Spread onto buttered cookie sheet. Bake in 350 degree oven for 30 minutes . Let cool before eating.

Debbie Harman

Millet Breakfast Cereal

1 C. millet
1 t. salt
1 C. raisins
1 qt. water
1 C. chopped apples
honey and milk

Put the millet, water and salt in a saucepan. Simmer for 45 min. or until millet is soft. Add apples and raisins and cook 15 min. longer. Serve with honey or sugar and milk.

Vicki Tate

Grapenuts Cereal

3 C. wheat flour
1 C. brown sugar
1 C. sour milk or buttermilk
2 t. soda
1 t. salt

Mix thoroughly. Bake 300 degrees until golden brown. Crumble and dry.

Jill Hansen

Coconut - Oatmeal Cold Cereal

3 C. white or wheat flour
2 C. coconut
3 T. water
2 C. quick oatmeal
3/4 C. white or brown sugar
5 T. oil

Mix well in large mixing bowl. Spread on 2 large cookie sheets. Bake about 45 min. at 350 degrees. Stir occasionally to crumple into size pieces you desire. You can also add raisins, dried fruits, or nuts.

Jill Hansen

Honey Coconut Granola

4 C. rolled grain
1 C. whole wheat flour
3/4 C. honey
1 C. coconut
1/4 C. cooking oil
1 t. vanilla

Thoroughly mix together. Spread on a cookie sheet , bake at 300 degrees for 20 minutes stirring occasionally while baking.

After cooking add any or all of the following:

1/4 C. sunflower seeds
1/4 C. coarsely chopped almonds or pecans.
1/4 C. sesame seeds
1/2 C. raisins or other dried fruit

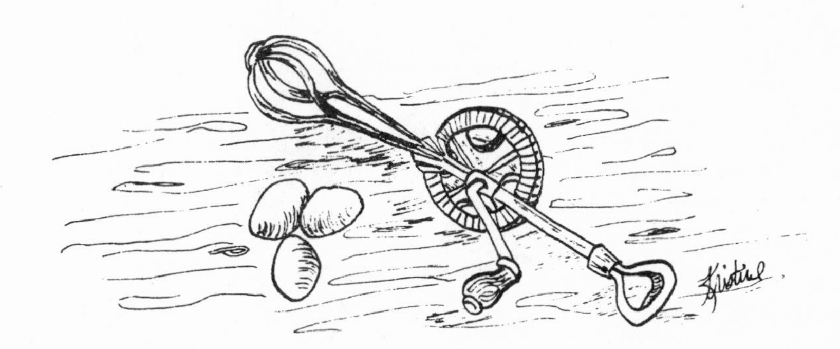

Dutch Apple Oatmeal

1 1/4 C. rolled oats
1 1/4 C. water
1-2 apples
cinnamon
1 1/2 C. milk
1/4 t. salt
vanilla

Place oats, milk, water and salt in a sauce pan and let stand 10 minutes. Bring to a boil. Reduce heat and simmer 10-15 minutes until oatmeal is thick and creamy. Stir in apples and honey, vanilla and cinnamon.

Vicki Tate

Instant Oatmeal Packets
(Good for camping)

Blend **1/2 C. of oats** until powdery.
Into each of 10 zip-lock sandwich bags combine:
1/4 C. regular oats
1/4 tsp. salt
2 T. powdered oats

To use: empty packet into bowl and add **1/2 C. boiling water** and stir until thick.

Variations To each packet add:

Apple-Cinnamon: **1 T. sugar, 1/4 tsp. cinnamon, 2 T. chopped dried apples.**
Cinnamon -Spice: **1 T. sugar, 1/4 tsp. cinnamon, 1/8 tsp. nutmeg.**
Raisins and Brown sugar: **1 T. packed brown sugar, 1 T. raisins.**
Wheat Germ: **2 T. any kind wheat germ.**

Farmhouse Breakfast Rice and Raisins

1 C. rice
1/4 - 1/2 raisins
4 C. milk

Combine rice, milk and raisins in saucepan. Simmer 10 - 15 min., till rice is tender.

Featherlight Buttermilk Wheat Pancakes

1 1/2 C. buttermilk
1 1/2 C. wheat flour
2 T. oil
2 t. vanilla
3 eggs, seperated
1/2 t. salt
1 T. baking powder

Combine buttermilk and egg yolks in bowl and beat until foamy. Add remaining ingredients, except egg whites, stir till smooth. Beat egg whites until stiff. Fold egg whites into batter. Delicious!

Vicki Tate

Whole Wheat Hot cakes

1 C. wheat flour(may use part white flour)
1 level t. baking powder
2 T. oil
2 T. sugar
1/4 t. salt
1 scant C. milk
1 large egg, beaten lightly

Sift dry ingredients. Combine milk and oil and add to dry ingredients. Then stir until moist but not too much. A cast iron skillet at medium heat is ideal. Use what you have.

Syrup: 1 C water, 1 C. sugar, heat to boiling and add some maple flavoring, about 2 t. approx.

Charles M. Larsen

Pancakes from a Mix

Master Mix on page 73

2 C. Wheat Quick Biscuit Mix
1 C. Water
2 Eggs, beaten

Combine all ingredients and stir until moistened. Cook on hot griddle.

Lisa Conn

French Toast

6 slices of bread
6 T. dried egg mix
1 C milk
1/4 t. salt

Beat the egg mix with the milk. Add salt. Dip each slice of bread in egg mixture and fry. It is good to add a bit of cinnamon and sugar to the egg mixture. Top the French toast with syrup or fresh fruit.

Fabulous French Toast

9 slices of dry white bread
3/4 C pancake mix
2 eggs
1/2 t. vanilla
3/4 C milk

Honey Butter

1/2 C honey
1/2 C softened margarine

Cut the bread in half. Place mix, vanilla, eggs, and milk in a bowl. Beat together until smooth. Dip bread in batter and fry until golden brown. Serve with honey butter.

Cherie Harmon

German Pancakes

1/2 C. milk
3 eggs
3 T. butter
1/2 C. flour
1/4 t. salt

Combine milk, flour, eggs, salt and blend. Melt butter in 8" x 8" pan in oven until hot and sizzling. Pour batter into hot pan. Bake 450 degrees for 25 min.

Anna Jean Hedelius

Buttermilk Pancakes

2 C. whole wheat flour
1/2 t. salt
2 T. baking powder
1 C. buttermilk
1/3 C. oil
2 T. honey or sugar
1 t. soda
1 1/2 C. milk
2 eggs

Mix and drop spoonfuls on hot griddle

Louise P. Eddy

Griddle Cakes
(White Flour)

1 egg
1 T. oil
2 t. baking powder
1 T. sugar
1 C. milk
1 C. flour
1/2 t. salt

Sift dry ingredients together. Combine egg, milk, oil and stir into flour mixture. Cook on grill or skillet till top is bubbly and bottom browned. Try adding apples to the batter.

Vicki Tate

"Olden Days " Wheat Crackers

4 C. wheat flour
2 t. salt
1/3 C. oil
1 T. yeast
2/3 C. powdered milk
1 1/2 C. warm water
1 T. honey

Mix dry ingredients. Dissolve yeast and honey in 1 C. warm water. Add this to dry ingredients. Add oil. Add remaining water. Let rise 1 hour. Knead. Divide dough into fourths. Roll each piecee as thin as possible on floured surface. Place on cookie sheet and bake at 350 degrees for 6 min. Turn crackers over and bake 2-3 minutes more. Cool and break into pieces.

Graham Crackers

1/2 c. evaporated milk
2 T. lemon juice or vinegar
1/2 C. honey
2 t. vanilla
1 t. salt
6 C. whole wheat flour (approx)
1/2 C. water
1 C. packed brown sugar
1 C. vegetable oil
2 eggs. beaten lightly
1 t. soda

Mix together milk, water and lemon juice. In separate bowl beat well, sugar, honey, oil, vanilla and eggs. Combine mixtures with dry ingredients. Divide in 4 equal parts. Place each on a greased and floured cookie sheet and roll to about 1/8" thick. Prick with fork. Bake at 375 degrees for about 15 minutes or until light brown. Remove from oven and cut in squares immediately.

LeAnne Beal

Wheat Thins

A thin batter:
1 C. wheat flour
1 C. water
1/2 t. salt

Mix until free of lumps. Grease cookie sheets. Spread 1/2 C. batter on entire sheet (do several sheets). Season with celery, onion, or garlic salt. Bake at 350 degrees 10 minutes. Remove and break up to chip size pieces.

Vicki Tate

Corn Chips
(Dried foods)

1/2 C. dehyd. corn
1 1/2 C. boiling water
1/2 1 C. dehyd cheese

Simmer corn and water for 25-30 minutes. When water is absorbed, pour corn in blender and puree. Add cheese and puree some more. Spread in thin layer on buttered cookie sheet. Sprinkle lightly with seasoning salt of your choice. Bake at 250 degrees until partially dried. Score with knife, so will fall into chips when dry. Continue baking until dry but not brown. It will dry more when cool. Lift off pan and cool.

Soda Crackers

3 T. shortening
2 T. butter
3 1/2 C. flour (approx.)
1 tsp. baking powder
3/4 tsp. salt
1/4 tsp. sugar
1 egg white, beaten to loose froth
1 C. whole milk

Melt the shortening and butter together; let cool. Sift 1 3/4 cups of the flour into a large mixing bowl along with the baking powder, salt and sugar; make a well in the center and pour in the shortening and butter; stir to combine, then stir in the egg white. Add the milk and beat vigorously to smooth out lumps, then gradually sift in enough flour to make a dough that will be stiff enough to roll out. Form the dough into a ball; knead it for 10 minutes on a floured surface until it is smooth and satiny. Divide into two balls, wrap each in wax paper and let them rest at room temperature for 20 to 30 minutes.

On a floured surface, roll dough ball out as thinly as possible. Cut the dough into 2-inch squares using a pizza cutter or sharp knife; save the scraps. Place the dough squares just slightly apart from each other on an ungreased cookie sheet. Sprinkle the squares with salt to taste and pat them to embed the salt. Prick with a fork. Bake at 375° for 15 to 20 minutes.

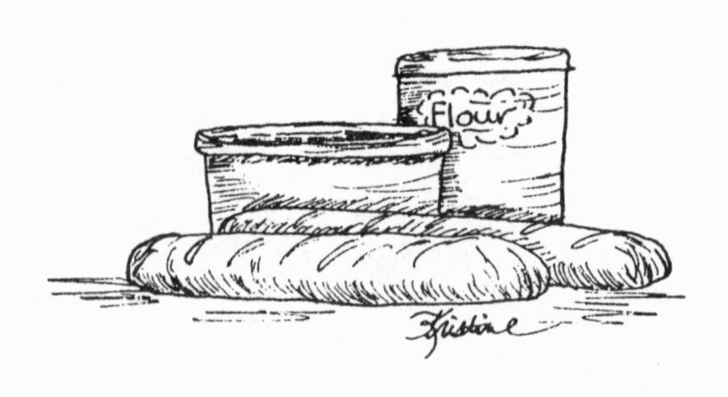

Croutons

Let **bread** dry out at room temperature for a few days or put in the oven at 250°- 300°until dry and crisp. Cut into 1-inch cubes. Toss with **oil, garlic powder, and herbs of your choice. Parmesan cheese** is good, too. Bake at 400° . until golden and crisp, about 10 minutes. Cool and store until ready to use.

Soft Pretzels

1 C. lukewarm (110 F.) water
1/4 C. sugar
2 T. shortening, softened
3 1/4 to 3 3/4 C. flour
Coarse or kosher salt for sprinkling
1 envelope dry yeast
1 tsp. salt
1 large egg, room temp.
1 T. water

In a warm mixing bowl, pour the water and sprinkle the yeast on top. Wait a few minutes then stir to dissolve. Dissolve the sugar, salt, and the shortening. Separate the egg; combine half of the yolk with the 1 T. water; cover and reserve for later. Mix the egg white and the other half of the yolk into the yeast mixture. Vigorously beat in 2 C. of the flour until smooth; then beat in enough additional flour to make a stiff dough. Cover the mixing bowl tightly and refrigerate the dough for 4 -24 hours.

Place the dough on a lightly floured board and cut it into 16 equal pieces. Roll 1 piece into a 20-inch rope then shape that rope into a pretzel (see box below) using a few drops of water as glue. Place the formed pretzel on a lightly greased cookie sheet and repeat process. Brush all pretzel tops with the reserved yolk-and-water mixture then sprinkle them with the coarse salt. Cover the pretzels very loosely and let them rise until doubled in bulk. Bake at 400° . for about 15 minutes or until amber and done.

Tortilla Chips

1 dozen corn tortillas | **oil for frying**

Cut tortillas into triangles (like a pie) and fry a few at a time in a pan of hot oil. Salt to taste.

Variation Try frying flour tortillas the same way and sprinkle with cinnamon and sugar mixture.

Wheat-Rye Thins

2 C. whole wheat flour
1 1/4 C. rye flour
1/3 C. oil
1 C. water
1 t. salt

Mix dry ingredients. Blend oil and water in blender, then stir into dry ingredients. Knead slightly. Divide into 2 ball. On bottom of cookie sheet, roll one ball to edges until uniformly thin. Mark cracker squares with a knife, do not cut all the way through the dough. Prick with a fork. Spinkle with salt, seasoned salt, onion or garlic salt, or herbs, as desired. Bake at 350° for 20 minutes, until golden and crisp. Let cool, then break apart.

Beans and Rice
BEANS
RICE

Cooking guidelines for beans

Dry Beans	Cooking time	Pressure Cook (15 lbs)
Black beans	2 hours	5 min
Black-eyed peas	1/2 hour	Do not pressure cook
Great Northern beans	1 1/2 hours	3 min.
Kidney beans	2 hours	3 min.
Lentils	1/2 hour	Do not pressure cook
Lima beans, large	1 hour	3 min.
Pinto beans	2 hours	10 min.
Navy beans	2 hours	7 min.
Split peas	1/2 hour	Do not pressure cook

Hints

1. Rinse all beans and legumes in cold water. Remove all dirt, rocks, or bad beans.

2. Soak the beans in 3 times the amount of water as beans. They can be soaked overnight. Lentils and split peas do not need to be soaked.

3. Quick soaking method: Boil the beans in water for 2 minutes, remove from heat, cover, and let stand for 1 hour.

4. Add 1 t. salt per cup of beans and use a large enough pan because the beans double in volume.

5. Add 1/8 t. baking soda and 1 T. cooking oil to each cup of beans while soaking. This will shorten the cooking time and decrease foaming.

6. Add meat, onions, celery, and herbs during cooking to add more flavor. Add tomatoes, catsup, vinegar and other acid foods after the beans are tender. The acid prevents softening of the beans.

7. Cooked beans freeze well and will keep up to 6 months in the freezer.

Dried beans were a staple to the early settlers. They stored well, were versatile and provided many hearty meals.

Mormon Baked Beans

2 C. small white beans
6 C. water
2 T. dehydrated onion
1/4 C. oil
1/4 C. brown sugar
3 T. honey
1/4 t. dry mustard
1 1/2 t. salt
1/8 t. pepper
1/2 C. bacon or bacon bits (optional)

Soak beans overnight. Simmer over low heat 1-2 hours until tender. Drain, reserving liquid. Add onions to beans and put into a 2 quart casserole dish. Stir together oil, sugar, honey, mustard, salt, pepper and 1 cup of reserved liquid. Pour over beans and stir gently. Add enough of remaining liquid to almost cover beans. Bake at 300 3-4 hours. Stir in bacon or bacon bits last 30 minutes.

Vicki Tate

Brother Brigham's Honey Beans

1 1/2 lb. white beans, dried
3 T. margarine
2/3 C. honey
salt to taste

Soak beans. Cook until tender. There should be 1 cup liquid remaining in beans. Add remaining ingredients, mix and simmer until beans are just juicy. (15 - 20 minutes) Note: You can substitute honey for molasses or part of brown sugar in regular baked bean recipes.

Cherie Harmon

Cowpolk Beans

2 lbs. pinto beans
2 lbs. ham hock
2 onions, chopped
4 T. sugar
2 green chilies
1 can tomato paste
salt and pepper to taste

Wash and soak beans overnight. Drain. Place in a dutch oven or heavy pan. Cover with water. Add all remaining ingredients and cook until tender. Add more water if needed. Salt and pepper to taste. Cook in oven for 2 hours or until done.

Sheep Camp Bean Loaf

1 1/2 C. cooked dried beans
1 onion, minced
1/4 C. celery
2 eggs, beaten
3/4 C. wheat bread crumbs
1/2 t. salt
2 t. melted margarine

Mix all ingredients together. Pat into loaf shape and sprinkle with paprika. Bake at 350 15 - 20 minutes. Serve with green pepper sauce.

Cheri Harmon

Baked Beans Western Style
(Dried Foods)

6 C. water
1 lb. red beans
1/4 C. dehydrated onions
2 C. tomato powder (reconstitute to desired thickness)
1 1/4 C. brown sugar
1 t. chili powder
3/4 C. bacon or bacon bits
salt and pepper to taste

Soak beans. Simmer with salt and onions until tender. Add remaining ingredients. Blend well. Place in casserole dish and bake 4 - 5 hours at 275 .

Beans (Refried)

1/4 C. bacon, cut up
1/4 C. onion, chopped
2 C. pinto beans
1 t. chili powder
1 t. garlic powder
1 t. salt

In crock pot on high temperature cook bacon and onions until tender. Add 2 quarts water and beans. Season with spices. Let cook several hours. Mash beans by hand or in blender. Great for Navajo Tacos.

Debbie Harmon

Beans Cooked in the Ground

Pioneer Recipe

Dig a hole about 18" square. Make a fire in the hole and let it burn down to hot coals. Place a pot of beans in the hole with plenty of water,in the pot, salt, pepper and 1 - 2 pieces of bacon. Cover tightly. Place coals and ashes around pot and cover with dirt. Cook 6-8 hours.

Elta Alder

Bean Burr-rr-ito

Make refried beans using pinto beans. Sprinkle cheese or dehydrated cheese over beans. Put 2 - 3 tablespoons bean mixture on a pliable tortilla and sprinkle in a little taco sauce. Roll the tortilla with the ends tucked in to keep filling in.

Refried Beans

Soak 2 cups pinto beans and 1 chopped onion overnight in 2 quarts water. Cook beans in water until tender (2 - 3 hours) checking to be sure beans don't dry out. Add 1 t. salt (or more to taste) and 1/4 cup oil. Beat with an electric mixer until smooth while still hot.

Sanpete Succotash

1 small can kidney beans or
or cooked dried beans
1 can whole kernel corn
6 T. margarine
1/2 t. chili powder
1/2 t. salt

Combine beans and corn in a casserole. Add margarine, salt and chili powder. Cover and bake in 375 oven for 25 minutes or until well heated.

Marilyn Ostler

Pinto Bean Casserole

2 C. cooked pinto beans
(adjust this to family size)
1 C. grated cheese
1 C. reconstituted cream mushroom soup
1/2 cup buttered whole wheat bread crumbs

Grease casserole, combine beans, cheese and soup. Bake 3/4 hour in 375 oven.

Betty Jenkins

Lima Beans and Short Ribs

1 C. dry lima beans
2 qts. water
2 lbs. short ribs
1/2 C. barley
1 T. salt
1 t. ginger
1/4 t. pepper
1/2 clove garlic

Soak lima beans in water for 6 hours. Place short ribs in kettle, add beans, water, barley, and seasonings. Bring to boil, cover, reduce heat and simmer 30 minutes. Cut meat into serving pieces and return to kettle. Bake in oven for 5 hours, covered.

Macaroni Soybean Loaf

4 T. margarine
4 T. flour
1/2 t. salt
pepper to taste
2 C. milk
1 C. grated cheese
3 C. cooked soybeans
1 C. cooked elbow macaroni
1 T. onion

Make a white sauce of the margarine, flour, salt, pepper, and milk. Add cheese and stir until melted. Combine with other ingredients and put in casserole. Bake at 350 for 40 minutes.

Mary Fehlberg

13 Bean Soup

2 C 13 bean soup mix
3 quarts water
1 ham bone
1 med onion
2 sliced carrots
2 sliced celery stalks
dash pepper
1/2 t. salt

Heat the beans and water to boiling. Boil for a few minutes. Remove from heat and cover. Let this stand for about 1 hour. Add the ham bone and simmer until the beans are tender, about another 2 hours. Stir in all remaining ingredients and cover. Simmer for 1 more hour. Remove the ham bone and trim all the meat from the bone. Stir meat into the soup.

Vicki Tate

Lentil Soup

2 C Lentils
2 chopped celery stalks
4 strips crisp bacon crumbled
1 chopped onion
2 T. spinach flakes

Sort and wash lentils. Put them in a pan or crockpot and cover them with water. Watch them so they don't go dry. Continue to add water if needed. Add the remaining ingredients and simmer until lentils are very soft. Always serve beans with a slice of whole wheat bread and a glass of milk.

Betty Jenkins

Bean and Bacon Soup

4 C white beans
1/4 lb bacon
2 C shredded carrots
2 t. garlic salt
1 diced onion
8 C water
1/2 cube butter or marg.
salt and pepper to taste

Boil the beans in water for 15 minutes. Cover and simmer until tender. About 2-3 hours. Add salt, pepper, bacon, carrots, and onion. Simmer for another hour. Just before serving add butter.

Robert Fowles

Lima Bean Soup

3 C Lima beans
3 T dried onions
1 t. dried parsley
4 chicken bouillon cubes
1 T. butter
1 t. salt
water as needed

Soak the beans overnight. Cook them for 3-4 hours . After 1 hour of cooking, add all other ingredients.

Louise P. Eddy

Crockpot Bean Soup

3 C any dried beans
1 med onion
1/4 t. garlic powder
1/4 C dried soup blend
1/4 t. savory seasoning
2 stalks chopped celery

Sort and wash the beans. Put them in a crockpot or kettle and add water to fill the crockpot about 2/3 full. Add remaining ingredients, turn the crockpot on high and let it simmer all day. If you are cooking in a kettle, simmer for about 3-4 hours. Check occasionally to see if more water is needed. Do not salt beans until they are soft.

Betty Jenkins

Tasty Legume Soup

1/2 pkg. onion soup mix
1/3 C rice
1/3 C lentils
1 lb. hamburger
1/3 C split peas
1/8 C barley
7-8 C water
1/3 C macaroni

Brown the hamburger. Combine all ingredients except macaroni and simmer for 3 hours. Add the macaroni or other pasta noodles about 30 minutes before serving.

Connie Gardner

Split Pea Soup

2 1/4 C green split peas
1 1/2 C chopped ham
1 1/2 C diced onion
1 t. salt
1/4 t. marjoram
1/2 t. pepper
1 C diced celery
1 C diced carrots

Add peas to 2 quarts warm water. Simmer 40 minutes. Add remaining ingredients. Simmer another 20-30 minutes

Judy Anderson

Bean, Vegetable, and Dumpling Soup

Soup

1 (15 1/2 oz.) can red kidney beans
1 (15 oz) can black beans, pinto or great northern
3 cups of water
1 (14 1/2 oz) can stewed tomatoes
1 can whole kernel corn
2 carrots sliced
1 large onion chopped
1 (4 oz) can chopped green chili peppers
2 T. beef or chicken bouillon cubes
1 1/2 t. chili powder
2 cloves garlic, minced

Dumplings

1/3 C flour
1/4 C yellow cornmeal
1 t. baking powder
salt and pepper to taste
1 beaten egg white
2 T. milk
1 T. cooking oil

Drain and rinse the canned beans. Combine beans, water, tomatoes, corn, carrots, onions, undrained chili peppers, bouillon cubes, chili powder and garlic. Cover and cook on low heat for 10 hours or on high heat for 4-5 hours.

For Dumplings, mix together egg white, milk, and oil. Add flour mixture and stir well until combined. Drop dumpling mixture by teaspoonfuls into boiling soup. Cover and cook for 30 minutes until done.

Peggy Layton

Smoky Flavored Split Pea Soup

8 C cold water
2 C green split peas
3/4 C diced celery
3/4 C diced carrots
3/4 C chopped onion
1 large bay leaf
1 t. salt
1/8 t. garlic powder
1/4 t. hickory smoke flavoring

Combine ingredients in a heavy pot . Cover. Bring to a boil . Reduce heat and simmer for 2 hours. Remove bay leaf. Serve.

Heather Mickelson

Rice and Beans Brazilian Style

2 C. pinto or kidney beans
6 C. water
Choose 2 - 4 of these vegetables:
- Potato
- Cabbage
- Pumpkin
- Okra
- Carrot

1/2 lb. ground beef
1/4 lb. bacon or sausage
2 cloves garlic
1 medium onion, chopped
1/2 green pepper
1 T. Worcestershire sauce
3 T. dehy. tomato powder
1 t. coriander
1 bay leaf
salt and pepper

Soak beans and cook until tender. In sauce pan cook choice of vegetables, cut in large pieces, until just tender. Sauté the meat and other ingredients in skillet. Simmer 30 minutes. Mix beans, vegetables and meat mixture. Heat together for about 2 minutes. Serve with rice.

Rice and Beans Carribean Style

2 C. pinto or kidney beans
1 T. salt
6 C. water

2 T. oil
1 clove garlic, crushed
2 green onions, chopped
1 large tomato, chopped
1/8 t. cloves
1 T. parsley flakes
1/4 t. pepper
2 C. rice
4 C. reserved bean juice

Soak beans. Cook until tender. Drain juice from beans, reserving for future use. Sauté garlic, onion, tomato and seasonings about 5 minutes. Add rice and reserved bean juice. Bring to a boil, reduce heat and simmer 20 - 25 minutes without stiriing.

Sheep Camp Lentils and Rice

2 large onions, sliced
1/4 C. oil
1 C. lentils
4 C. water or stock
1 1/2 t. salt
1/4 t. pepper
1/2 C. rice

Fry onions in oil until light brown. Set aside half the onions. Rinse lentils and pick over. Put in 3 quart pan, add water, bring to boil, cover and cook over low heat 20 minutes. Add rice, salt, pepper and other half of onions. Continue cooking until tender, but not mushy. About 25 minutes. Variation: use millet in place of rice. Add 1 cup mushrooms.

Granddad 's Brown Rice and Beans Supper

1 C. brown rice
2 1/4 c. water
1/2 t. salt
1 T. butter

Sauce:
3 8 oz. cans tomato sauce
1/2 C. salsa
1 can kidney beans
3 T. dried onions
2 green peppers
1/2 t. salt
1/4 t. Italian seasoning
1 T. butter
1/2 C. water

Cook rice for 45 minutes. Mix sauce and cook for 5 minutes. Top all with cheese (Monterey Jack).

Louise P. Eddy

One Pot Brown Rice and Vegetable Supper

4 C. short grain brown rice
6 C. cold water
2 t. chicken bouillon
1 C. chopped onion
2 medium carrots, chopped
2 stalks celery, chopped
1 t. salt
1/4 - 1/2 t. cayanne

Put all ingredients into a 6 quart pan. Bring to a boil on high and reduce heat to medium and cover pan. Cook about 45 minutes. For a more moist rice add more water.

Debbie Wade

Rice and Beans Italian Style

1/2 C. navy beans
1 onion
1 stalk celery
1 C. canned tomatoes (with juice)
1 t. salt
1/2 C. rice
2 T. oil
pinch hot red pepper
2 T. parmesan cheese

Soak beans. Simmer until tender. Brown onion and celery in oil and add to beans. Add tomatoes, salt and pepper. Add rice, cover tightly and cook, stirring often until rice is tender (about 20 minutes). When cooked, the mixture should have the consistency of a thick stew.

Jennie's Mexican Rice

1 1/2 C. white rice
1/4 C. oil
1/4 C. chopped onions
1 clove garlic

Brown rice in oil. Add onion and garlic. Sauté.

Add:
2 C. chicken broth
1 C. water
1/4 t. cummin
2 t. salt
1 t. pepper
1 t. garlic powder
1 can (8 oz.) tomato sauce

Bring to boil, then simmer 20 -30 minutes with lid on. Check periodically. Add more water if necessary but do not stir. Cook until rice is soft and liquid is gone. Serve with refried beans, enchiladas and shredded lettuce.

Connie Gardner

Chicken Rice Casserole

(One of President Kimball's favorites)

4 eggs, hard boiled, diced
2 cans cream of mushroom soup
1 C. celery, chopped
1 C. onion, chopped
3 C. liquid or broth
2 T. soy sauce
1 whole chicken, boiled, boned and cubed
1 1/2 cup uncooked rice

Prepare chicken, rice, eggs and set aside. Melt 4 tablespoons butter in skillet and sauté onion and celery until golden. In a large bowl mix chicken cubes, eggs, celery, onion mixture, soup and soy sauce and liquid. Stir until well blended. Butter 2 quart casserole dish. Place half of rice in an even bottom layer. Spoon in half of chicken mixture. Next, spread in the rest of rice, then remaining chicken mixture. Bake at 350 for 20 to 25 minutes.

Sister Smith's Rice Dinner

1/2 C. celery, chopped
1/2 C. onion, chopped
1 can cream of mushroom soup
1 T. cornstarch
1 t. onion powder
1 can tuna or chicken

Cook celery and onion in 3 1/2 cups water. Add soup. Combine cornstarch with 1/4 cup cold water. Add to soup mixture along with onion powder and tuna or chicken.. Heat through and serve over rice.

Judy Anderson

Old West Spanish Rice

1 C. rice
1 1/2 C. water
1 onion, chopped
few sprigs celantro
garlic salt
green peppers or jalepeno peppers (your choice)
1 can tomato sauce (8 oz.)
2 T. oil
1 C. green onions

Fry rice, onions, and peppers in oil until lightly browned. Add water, tomato sauce, celantro and salt. Let boil rapidly for 3 or 4 minutes; cover and simmer 15 minutes. Sprinkle green onions on and serve.

Robert Fowles

Country Baked Rice Pilaf Supper

(Fresh or Dried Foods)

1/2 C. butter
1 C. rice
1 large onion*
1 C. mushrooms*
1/4 C. green peppers*
2 C. chicken bouillon

Melt 1/4 cup butter in skillet. Sauté onion, mushrooms and green pepper until tender. Remove and set aside. In same skillet melt remaining butter. Add rice and stir until lightly browned. Stir in vegetables. Add bouillon. Pour into casserole dish and bake for 40 minutes at 350 .

Vicki Tate and Leslie Powell

Fried Rice

3 slices chopped ham or bacon
1 carrot
2 T. green onions
1/2 C. peas
pepper to taste
MSG
3 C. cooked rice
soy sauce to taste`

Fry bacon. Chop carrot into small pieces and brown in bacon grease. The grease must be very hot. Add rice, green onions and a handful of green peas. Stir constantly. Season with pepper and MSG or Accent. Pull rice to the side. Crack 2 eggs into center and fry. Sprinkle with soy sauce. Sprinkle bacon, ham or other meat and green onions on top.

Cathy Call

Round-Up Time Tuna and Rice

1/2 C. chopped onion
2 T. margarine
3 T. flour
1/2 C. milk
salt to taste
1 t. dry mustard
1 (7 oz.) can tuna
2 C. cooked rice
2 t. lemon juice
2 T. water
1 egg, slightly beaten
whole wheat flour

Sauté onions in margarine until soft. Blend in flour, add milk and seasonings. Cook until thickened; remove from heat. Add tuna, rice, and lemon juice. Chill. Form into patties. Mix water and egg. Dip tuna patties into egg mixture; roll in flour and chill about 15 minutes to dry. Fry in shallow fat until brown on each side. Serve with creamed peas and carrots. Serves 6.

Connie Gardner

Sunday Rice

1 lb sausage
1 lb ground beef
1 onion, chopped
1 green pepper, chopped
1/2 C. chopped celery
3 C. boiling water
2 pkg Noddle soup mix
1 C. uncooked rice

Brown sausage, ground beef, and onion. Dissolve soup mix in boiling water. Mix all the ingredients together and bake in a covered dish at 425° for 15 minutes. Turn oven down to 200° and bake for 3 1/2 hours. You may top with slivered almonds.

Meat Substitutes
FLOUR
BEANS

“Meat” from Wheat

(Gluten)

8 C. wheat flour
2 C. water

Combine and mix into ball. Knead, pound or beat for 10 - 15 minutes. Cover in cold water for 1 hour. Wash out starch by rinsing dough under hot water until dough is firm. Continue washing until the bran (feels like sand) has been washed out. Water will be almost clear. Let drain for 30 minutes. Flavor by using one of the following recipes.

Beef Flavoring for “Meat”from Wheat

2 T. beef bouillon
1 T. soy sauce
4 t. seasoned salt
1 t. pepper
1/2 C. onion

Mix ingredients and bring to a boil. Add gluten pieces and simmer 30 minutes.

See following recipes to make “ground beef ”, “steaks”, and “sausage”.

“Meat” from Wheat Ground Beef

(Gluten)

Simmer gluten in beef flavoring. Grind in meat grinder. Substitute in recipes calling for ground beef.

Campfire "Steaks"
(Gluten)

Gluten steaks
1 C. dehydrated mushrooms (reconstituted)
1/2 C. margarine
salt and pepper

Shape gluten into steak shapes and season with basic beef flavoring. Fry steaks in butter. Sauté mushrooms with steaks for 5 minutes. Serve.

Ranchhouse "Steaks"
(Gluten)

gluten steaks
1 egg, beaten
2 C. fine bread crumbs
1/2 C. margarine
6 slices bacon (or bacon bits)
1 can mushroom soup

Form gluten into steak shapes and season with basic beef flavoring. Fry bacon, crumble up into fine pieces. Return to pan. Brown steaks in butter with bacon. Pour mushroom soup over steaks. Heat and serve.

Log Cabin Stew
(Gluten)

5 C. gluten cut into stew cubes
4 - 5 celery stalks, chopped
2 C. potatoes, chopped
1 C. tomato sauce
1 chopped onion
2 C. carrots, chopped
1 C. tomato soup, juice or tomato powder (reconst.)

Cover stew cubes with water and add 3 - 4 boullion cubes. Boil 1 hour. Add remaining ingredients and simmer until vegetables are tender.

Mary Fehlberg

Wheat "Meat" Hash
(Gluten)

1 can (28 oz.) or 3 1/2 C. tomatoes
1 C. green peppers
1/2 C. chopped onions
1/2 C. uncooked rice
1/2 t. beef soup base
1/2 t. basil
Dash pepper
2 cups gluten ground beef
1 C. grated cheese

In a skillet, combine all ingredients except wheat meat and cheese. Simmer, covered, until rice is tender. Add wheat meat, top with cheese, and return to heat until cheese is melted. Makes 4 to 6 servings.

Mock Hamburger

1 C. wheat
1 can red kidney beans
(or any cooked beans)
1 egg
salt and pepper to taste

Cook wheat 1 hour. Blend wheat in blender then add beans and continue to blend. Add egg and salt and pepper. This can be molded into patties and fried or used in any recipe calling for hamburger.

Cherie Harmon

Wagon Train Bean Burgers

1 T. oil
1 medium onion, chopped
1 clove garlic, chopped
2 C. drained, mashed, cooked
pinto beans, kidney beans
1/2 t. oregano
1/2 C. wheat bread crumbs
2 egg yolks
2 T. milk
1 t. salt
1/4 t. pepper

Heat oil in skillet. Sauté onion and garlic until tender. Combine with beans, crumbs, eggs, milk, salt, pepper and oregano. Mix well. Shape into 4 burgers. Chill for 1 hour or longer. Sauté burgers until lightly browned. Serve with BBQ sauce.

Vicki Tate

Ranch House BBQ Beef
(Dried Foods)

2 C. beef flavored gluten, T.V.P. or dehydrated beef patty
1 C. dehydrated tomato powder
1/2 C. dehydrated onion (reconst.)
1/4 brown sugar
1/2 C. vinegar
1/2 C. worcestershire sauce
2 t. dry mustard
2 t. salt
6 C. water

In saucepan mix water and tomato powder until smooth. Add rest of the ingredients and cook over low heat, covered for 30 - 45 minutes. Stir frequently. Add water if thickens too much. Serve over rice or on hamburger buns.

Lisa Conn

Mock Salmon Loaf

2 C. ground soy beans
1 C. bread crumbs
1/2 C. milk
3 T. tuna fish
1 egg , beaten
1/4 C. sautéd onion
1/4 C. sautéd celery
2 T. sautéd green pepper
1/2 t. salt

Mix together. Bake at 350 for 45 minutes.

Mary Fehlberg

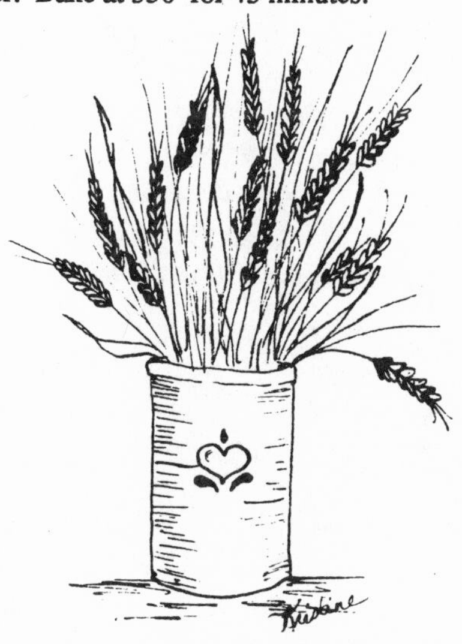

VegeMillet Burgers

1 C. millet
1 T. oil
1/8 t. cayenne
3 C. boiling water
1 t. sea salt
3/4 C. grated carrot
3/4 C. minced onion
1/2 minced parsley
1/2 C. wholewheat flour
1/4 C. soy flour
oil for frying

Place a large pot on medium heat and add oil. When oil is hot add millet and cayenne and stir for 2 - 3 minutes until millet gives off a nutlike fragrance. Add boiling salted water and reduce heat to simmer. Simmer, covered, for 35 - 40 minutes, adding vegetables 5 minutes before done. Add flours to millet and vegetables mixing well to avoid lumps. Let mixture sit for 15 - 20 minutes until cool enough to handle. Heat oil about 1/8 " deep in a frying pan on medium-low heat. Form millet mixture into thick patties with hands, or press into shape with a pancake turner. Fry for 3 - 4 minutes on each side or until crisp and lightly browned. Serve with soy sauce or your favorite sauce or gravy. Makes 18 - 20 burgers.

Jane Braithwaite

Wheat Taco Filling

1/2 C. wheat sprouts
1/2 C. steamed wheat (bulgar)
1/2 C. gluten
1/2 C. bread crumbs

Season with taco seasoning or Lipton's Onion Soup Mix. If you need more moisture, add 1/2 cup water with 1 beef bouillon cube.

Mary Fehlberg

Simple "Sausage"

(Gluten)

Gluten ground beef
1 T. flour
1/2 C. margarine
sausage seasoning

Season gluten ground beef with sausage seasoning to taste. Stir in egg and flour. Shape into patties or links and fry in margarine.

No Meat Sausages

2 C. cooked soybeans
1 C. cooked lima beans
1 C. cooked navy beans
2 t. salt
1/8 t. paprika
1 T. butter
1/2 t. sage
1/2 t. thyme
1/4 marjoram
1/4 t. savory
1 egg, beaten
2/3 C. milk
1 C. corn meal

Purée all beans. Add seasonings and mix well. Shape the bean mixture into sausage shapes. Combine milk and egg. Dip sausages into milk mixture, then roll in cornmeal. Put sausages onto cookie sheet and bake in 500 oven until browned on all sides. Turn during cooking.

Soy Bean Patties

2 C. soy beans, mashed
2 C. cooked brown rice
2 T. oil
1 onion, chopped
1 T. soy sauce
1 t. garlic salt
poultry season to taste

Mix all together and form into patties. Roll in whole wheat crumbs. Fry or bake until brown. Serve with brown gravy.

Brown Gravy:
2 T. oil
2 T. wheat flour
1 C. water
1 T. soy sauce

Cook until thick.

Mary Fehlberg

Wheat "Meat" Stroganoff
(Gluten)

1 medium onion, chopped
3 T. margarine
2 T. flour
1 t. soup base
1 clove garlic
3 C. cooked rice or noodles
1/8 t. pepper
1 can (4 oz.) mushrooms
1 C. milk
1 cup sour cream or yogurt
2 cups gluten ground beef

In a large skillet, sauté onion in margarine until tender. Stir in flour, soup base, garlic, pepper, and mushrooms. Stir in milk and bring to a boil; reduce heat and simmer uncovered for 10 minutes. Stir in sour cream and wheat meat. Serve over rice or noodles. Makes 4 to 6 servings.

Cabbage Rolls
(Gluten)

2 C. water
1 T. vinegar
1 t. salt
1 medium cabbage
1 medium onion
2 T. butter
2 C. cooked rice
2 C. gluten ground beef
2 cans (10 oz.) tomato soup
1 soup can water

Bring water to a boil in a 4 quart saucepan. Add vinegar and salt. Core cabbage and add to boiling liquid. Simmer until leaves become soft and are easy to remove. As leaves become tender, remove them from boiling water and cool.

Sauté onion in butter until tender. Add rice, gluten ground beef, and 1 can tomato soup. Spoon wheat meat mixture in cabbage leaves and fold over. Stack in a pan or casserole dish. Mix 1 can tomato soup with 1 can water and pour over cabbage rolls. Bake uncovered at 350 for 45 minutes. Makes 6 to 8 servings.

Traditional Hamburger Pie
(Dried Foods)

1 dehydrated beef pattie or
  or 1 C. ground beef gluten
1 can condensed tomato soup
1 can green beans or 1/2 C.
  dehydrated green beans, recon.
1/4 C. dehydrated onions
1/2 C. dehydrated potato
  flakes
2 1/2 C. boiling water
1/4 C. powdered milk
2 T. margarine powder
salt and pepper to taste

Reconstitute meat and vegetables. Brown ground beef with onions. Add green beans and tomato soup. Mix well. Salt and pepper to taste. Pour into casserole dish and top with mashed potatoes (reconstituted potato flakes). Bake 30 minutes at 350 .

Bunkhouse Hamburger Casserole
(Dried Foods)

1 can cream of chicken or cream
  of mushroom soup
2 T. soy sauce
3 C. cooked rice
hamburger or T.V.P.

Cook cream of mushroom soup and add soy sauce, until well mixed. Brown hamburger or reconstitute T.V.P. Mix all with rice. Bake at 350 for 20 minutes. Serves 6.

Lorna Fowles

Main Dishes
BEANS

Pork was a staple of early pioneer families. Most families had a pig which was butchered in the fall The fat was rendered into lard. Hams and bacon were cured and smoked to preserve them. Usually the meat was wrapped in a clean flour sack and paper and buried in the cellar where it would keep all winter.

Salt Creek Corned Beef Casserole

1 C. chopped celery
1 chopped onion
3 C. cooked noodles
1 can corned beef
1 T. worcestershire sauce
2 C. corn
2 cans cream soup (any kind)
2 T. margarine
1 chopped green pepper
1 1/2 C. water

Sauté onions, celery and green pepper in margarine until tender. Mix remaining ingredients and fold all carefully into noodles. Bake at 350 until bubbly.

Connie Gardner

Papa's Favorite Beef Hash
(Dried Foods)

1 can corned beef
2 C. dehydrated diced potatoes
1 C. dehydrated onions
1/4 C. powdered milk
2 T. flour
1 1/2 t. salt
1/4 t. pepper
5 C. water
1/4 C. oil

Combine potatoes, onions, milk, flour, salt and pepper and water in large saucepan. Boil for 10 minutes or until liquid is mostly absorbed. Heat oil in skillet. Add potato mixture and corned beef. Cook over low heat, turning frequently until potatoes are tender and brown. About 20 minutes.

Corned Beef and Cabbage
(Dried Foods)

4 C. dehydrated cabbage
4 C. boiling water
1 can corned beef
3 C. medium white sauce
1/2 C. dehydrated cheese

Pour boiling water over cabbage. Simmer until water is absorbed (about 15 minutes). Break corned beef into pieces and mix into cabbage. Pour into casserole. Pour on white sauce. Sprinkle cheese on top. Bake 30 minutes at 325 .

Marilyn Ostler

Country Style Corned Beef Hash

Potatoes, diced
1 can corned beef
onions, chopped

Fry potatoes and onions until tender. Add 1 can corned beef and cook until heated.

Marilyn Ostler

Corned Beef on Toast

1 can corned beef
6 T. margarine
1 1/2 T. dehydrated onions
6 T. flour
4 C. milk

Sauté onions in margarine. Add flour and milk, cooking until thickened. Add beef. Remove from heat. Serve over hot, buttered toast.

Mary Cowan

Patriarch Potatoes and Ham
(Dried Foods)

1 1/2 C. dehydrated potatoes
2 C. milk
1/4 C. margarine
1/4 C. flour
1/2 C. dehydrated cheese
1 MRE (Meals Ready to Eat) ham slice or 1/2 C. ham T.V.P
salt and pepper to taste

Reconstitute potatoes. Combine milk, flour, margarine, salt and pepper. Simmer until thickened. Layer potatoes, ham and cheese. Pour sauce over top. Bake at 325 40 - 45 minutes.

Ranchhand Jack's Skillet Dinner
(Dried Foods)

2 C. cooked whole wheat, cracked wheat or rice
1 dehydrated beef patty, 1 C. ground beef gluten or 1 C. mock hamburger
1 C. tomato powder
2 T. dehydrated gr. peppers
1/4 C. dehydrated onion
2 T. dehydrated celery
salt and pepper to taste

Reconstitute all dehydrated vegetables and beef. Brown meat. Add sauce and rest of ingredients. Simmer 20 - 25 minutes.

Skillet Potato Dinner
(Fresh or Dried Foods)

Reconstitute 2 cups dehydrated sliced or diced potatoes. Drain. Heat margarine and fry potatoes with choice of the following:

1- dehydrated onions
2- dehydrated broccoli
3- canned, stewed tomatoes
4- dehydrated cheese
5- chili
6- meat of choice: bacon bits, sausage gluten, canned chicken, etc.

Season to taste.

Chicken a-la-King
(Dried Foods)

1 1/2 C. chicken or chicken T.V.P.
2/3 C. butter or margarine powder
1 T. chicken bouillon
1/4 C. dehydrated onion
1/4 C. dehydrated mushrooms
2 T. dehydrated green peppers
1/3 C. oil
1 1/3 C. milk
1/2 t. salt
1/8 t. tumeric
1/3 C. flour
1 1/3 C. water

Cut chicken into small chunks or rehydrate T.V.P. Heat oil in skillet and sauté reconstituted vegetables. Blend in flour, margarine powder and spices. Dissolve bouillon in water and add to vegetable mixture. Add milk. Heat until it boils and boil one minute. Serve over rice.

Lisa Conn

Chicken Chow Mein
(Fresh or Dried Foods)

Brown lightly in oil:
1/4 C. onion*
1 C. mushrooms*

Add and simmer 15 minutes:
1 1/2 C. chopped cooked chicken (canned will work)
1 C. diced celery*
1 1/2 C. meat stock or water
2 T. soy sauce

Blend and stir into meat mixture:
1 1/2 T. cornstarch
3 T.water

***Dehydrated vegetables will work well.**

Cook until slightly thickened and clear. Serve over rice or noodles

Vegetable Bean, Teriyaki

1/2 C celery cut diagonally
1/2 C mushrooms sliced
2 C green beans
2 C sprouts
1/2 cup onion cut long ways
4 T. oil
2 C wax beans
2 T. soy sauce

Cook celery, onions, and mushroom until tender. Add other ingredients. Simmer until tender. Dried vegetables will work well if they are reconstituted first.

Mary Fehlberg

Aunt Kathryn's Meat Pies
(Empanadas)

Pie Crust

Filling:

1 1/2 lb. ground beef, browned
1 C. water
1 envelope onion soup
2 T. flour

Combine filling ingredients and simmer to thicken. Roll out pie crust to 6" circles. Put on meat filling, fold over and seal with water and fork. Bake 1/2 hour at 375 .

Options for filling:

1. Chopped cooked chicken or beef sauteed with onion, raisins, and green olives. Add 1/2 teaspoon cumin, 1 teaspoon chili powder.
2. Refried beans, cheese, and chopped chilies.
3. Tuna or salmon with mushrooms and chilies.
4. Cheese.
5. 3 oz. cream cheese blended with 3 tablespoons strawberry or apricot preserves. Sprinkle with sugar.

Vicki Tate

Beef Biscuit Pie

Beef Vegetables:
1 1/2 lb. stew meat
1/2 C. flour
2 t. salt
1/4 t. pepper
shortening
3 C. water
1 t. vinegar
1/2 t. sugar
pinch cloves
1 bay leaf
3 carrots, sliced
3 potatoes, diced
1 onion, diced

Cheese Biscuit Topping:
1 1/4 C. grated cheese
2 C. flour
4 t. baking powder
3 T. shortening (cut in)

Cook sliced up stew meat that has been coated with flour, salt and pepper mixture. Brown in shortening. Then add water, vinegar, sugar, cloves and bay leaf. Cook on low for 2 hours until tender. Add all vegetables and continue cooking until tender. Remove meat and vegetables, thicken the gravy and add back to meat and vegetables. Top with biscuit topping. Bake 20 minutes at 400 .

Cheese Biscuit Topping: Stir in 2/3 cup milk. Knead lightly and roll out like pastry dough. Cover casserole. Bake as directed above.

Farmhouse Chicken Pot Pie
(Fresh or Dried Foods)

2 C. canned chicken
1 T. chicken bouillon
2/3 C. flour
8 C. water
1/2 C. dehydrated carrots
1/2 C. dehydrated peas
1/4 C. dehydrated onions
1/2 t. pepper
pie crust pastry

Line casserole dish with 2/3 pie pastry. Cut chicken into small pieces. Combine chicken, bouillon, flour, carrots, peas, onions and pepper in large saucepan. Add water. Cook over medium heat until forms smooth thick gravy. Pour into pie crust lined casserole dish. Cover with remaining pie crust pastry. Bake 400 for 35 - 45 minutes until browned. *Variation: Instead of using pie crust, pour chicken combination into casserole dish and top with biscuits.

Lisa Conn

Fish was an important food for pioneer and Indian. In many communities as the settlers began to prepare for the winter, men were sent to the lakes and streams where they fished until they had caught enough to last through the winter. The fish was salted or smoked to preserve it.

Vicki Tate

Simple Salmon Patties

2 cans salmon
bread crumbs
oatmeal
2 - 3 eggs
onion salt

Mix well. Form patties and coat with flour. Add salt and pepper to taste. Fry on griddle until browned.

Elaine Westmoreland

Clam Fritters

1 C. flour
2 eggs
1 T. melted shortening
1/4 C. milk
1 can clams
salt and pepper

Mix and sift flour, salt and pepper. Add beaten eggs, milk and clam juice from can of clams. Mix until smooth. Add shortening and clams and mix well. Drop by tablespoons full into hot oil and fry until golden brown. Drain on paper towel. Serve with tarter sauce.

Etta Alder

City Folk Tuna Roll

Mix together:
1 1/2 C. flour
1 1/2 t. baking powder
1 t. sugar

Add:
1/4 C. oil
1/2 milk

Stir into soft ball and roll out. Spread with tuna filling. Roll as for jelly roll. Place on baking sheet. Slash roll diagonally into 6 pieces, cutting half way through roll. Bake at 450 for 20 minutes.

Tuna Filling:
1 can tuna, undrained
1/2 C. celery*
2 hard boiled eggs, chopped
2 T. onion*
2 T. salad dressing
1/2 t. salt
1/2 t. pepper

***Dehydrated works well**

Mix all together and spread on dough as directed above.

Cherie Harmon

Tasty Tuna Patties

1 can tuna
2 C. mashed or grated cooked potatoes
1 egg
2 t. lemon juice

Mix ingredients together and fry on a skillet as you would pancakes.

Judy Anderson

Chile Relleno Casserole

1 lb Montery Jack
1 lb Sharp Cheddar Cheese
1 can diced chiles

1 C. milk
1 C. flour
6 eggs

Grate cheeses and mix with the chiles, place in a well greased 9 x 13 pan . Blend together the milk, flour, and eggs. Pour over the cheese mixture and stir slightly. Bake at 375° for 45 minutes. Serve with salsa and sour cream.

South of the Border Enchiladas

Brown:
3/4 lb. ground beef
1 medium onion

Stir in:
2 C. refried beans
1 t. salt
1/8 t. garlic powder
(Heat until bubbly)

12 tortillas

Sauce:
1 medium onion, sauted
3 1/2 C. tomato sauce
1/2 t. garlic powder
1 - 2 T. chili powder
1/4 t. oregano
1 t. salt

Heat tortillas in hot oil to soften. Drain.

Pour 1/2 sauce into baking dish. Fill tortillas with 1/3 cup bean/beef filling and roll to enclose filling. Place in baking dish. Pour remaining sauce over filled tortillas. Cover with cheddar cheese or reconstituted dehydrated cheese. Bake at 350 for 15 - 20 minutes.

Option: Omit meat. Sprinkle cheese over refried beans before rolling up tortillas.

Creamed Corn Supper
(Pioneer Recipe)

10-12 pieces bacon
1/2 C. onion
1/2 C. flour
8 C. milk
2 t. salt
1/2 t. pepper
1/2 t. celery salt (added since)
2 C. corn

Cook bacon. Add onion and cook. Add flour to form a ball. Then whip in the milk until it forms a thick and smooth sauce. Add the corn and all other ingredients to taste.

Nachos

1 dozen corn tortillas
1 C. refried beans
1/2 C. dehydrated cheese
6 jalapeno peppers

Cut tortillas into quarters. Heat 1" oil in skillet. Fry tortillas until crisp. Drain on paper towel. Top with beans, reconstituted cheese and peppers.

Lisa Conn

Reservation Tamales

Meat from boiled chicken
2 cloves garlic or 1 onion
1/2 t. cayenne
2 t. salt
2 C. corn meal
2 - 3 small red peppers
corn husks

Chop the chicken; season with the cayenne pepper, garlic or onion and salt. Form the meat into little rolls about 2" long and 3/4" in diameter. Pour boiling water over the meal and stir. Use enough water to make a thick paste. Take a heaping tablespoon of the paste, pat it out flat and wrap a roll of chicken in it. Wrap each roll in corn husks which have been softened by immersing in hot water. Tie the husks with a piece of string close to each end of the roll. Cut off ends of corn husks. Cover the rolls with broth from chicken, or with boiling salted water. Add 2 - 3 red peppers and boil 15 minutes. Other meat may be used.

Mary Fehlberg

Huevos Rancheros

4 corn tortillas
2 green peppers
2-3 t. chili pepper to taste
2 tomatoes
1 onion
2 eggs per person

Heat tortillas. Saute vegetables until they begin to loose their shape. Add the seasonings. Fry 2 eggs per person. To serve place the eggs on the tortilla, cover with sauce and serve.

Sheepherder Tamale Pie

6 C. chili
2 C. cornmeal
5 C. water
1 t. salt

Combine water, corn meal and salt in double boiler. Cook covered for 30 minutes, stirring occasionally until quite stiff. Line bottom of casserole with 1/2 of the cornmeal. Pour prepared chili into casserole. Spread rest of cornmeal on top. Bake 30 minutes at 350 .

Mexican Vegetable Rice

1 T. cooking oil
1 onion, chopped
1 clove garlic, minced
1 1/2 C. uncooked rice
1/2 t. salt 1 C. chopped tomato
1/2 t. chili powder
3 chicken bouillon cubes
3 C. boiling water
1 carrot, shredded
1 (10) pkg. frozen peas

Heat oil, add onion, garlic and rice. Stir and cook until rice is opaque. Add salt and chili powder. Add water and bouillon cube. Add carrots into rice mixture. Bring to a boil. Cover and simmer 20 minutes. Add peas and tomatoes. Cook over low heat till peas are tender.

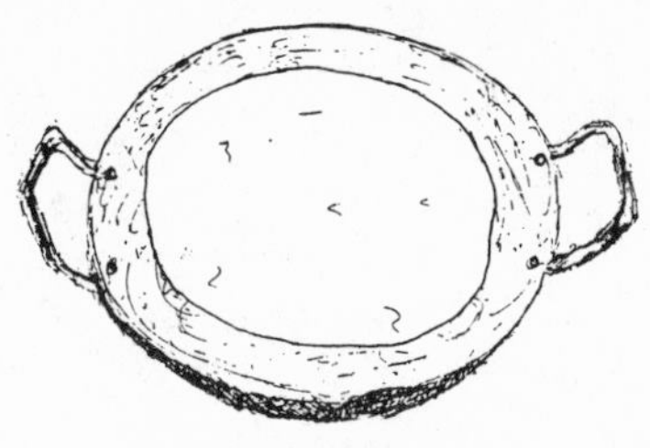

Enchilada Casserole

1 1/2 lb hamburger
1 onion, chopped
1 1/2 t. cumin
2 garlic cloves, crushed
4 t. chili powder
1 C. water
1 can corn
2 C. salsa
12 corn tortillas
1 C. sour cream
1 lb Monterey Jack
1 can olives

Brown meat and onion. Add next six ingredients. Simmer 5 minutes, uncovered. Pour 1/2 c. salsa into casserole pan. Arrange 6 tortillas to cover bottom. Add another 1/2 c. salsa on tortillas. spoon beef mixture on top of tortillas, top with corn, then olives. Top with sour cream and half the cheese. Arrange remaining tortillas overlapping slighty on cheese. Spread remaining sauce. Top with remaining cheese. Bake 40 minutes, covered: then uncover for 5 minutes.

Injun Corn Casserole
(Dried Foods)

1 C. dehydrated sweet corn
1/4 C. dehydrated onions
1 C. tomato powder
3 T. dehydrated green peppers
1 C. dry bread crumbs
2 T. dehydrated cheese
1 T. shortening or margarine

Reconstitute vegetables. Add seasonings. Place in casserole dish. Dot with margarine or shortening. Sprinkle with cheese and crumbs. Bake at 375 for 30 - 35 minutes.

Egg Noodles

2 C. flour
4 eggs
1 t. salt

Whip eggs until foamy, add salt and flour. Mix thoroughly. Divide dough into 4 equal parts. Roll dough into paper thin rectangles on a well floured surface. Cut into strips or squares depending on what you will be using the noodles for. Drop into boiling salted water (1 T. salt per 2 Qts water. This will cook half this dough.) until tender, 12-15 minutes. You can put them directly into boiling soup.

Homemade Pasta

2 C. wheat flour*
1 t. salt
2 eggs
1 egg white
1 T. oil
water
sifted flour

***white flour may be used**

Make a pile of 2 cups flour and the salt on a board. Make a well in the center and add the eggs, egg white and oil. With the fingers gradually draw the flour into the wet ingredients, adding drops of water as it seems necessary to form the mixture into a ball of dough. Knead the dough, using a minimum of flour on the board, until the dough is smooth and elastic, at least 10 minutes. Cover and let rest for 10 minutes. Divide dough in half and roll out each half until very thin. Dust lightly with flour and let dry for 10 minutes. Gently roll each rectangle of dough into a jelly roll shape and cut into 1/4 to 1/2 inch widths. Unfold bundles and set aside to dry.

Ranch Hand Macaroni and Cheese
(Dried Foods)

4 C. elbow macaroni
1 1/2 C. dehydrated cheese
2/3 C. powdered milk
1/4 C. cornstarch
1 T. dehydrated onion
3 C. water
1 t. salt
1/2 t. pepper

Cook macaroni until tender. Drain. Combine powdered cheese, milk, onion, salt, pepper, cornstarch and water in saucepan. Cook over medium heat until it boils and is smooth. Stir constantly. Boil 1 minute. Pour sauce over macroni and serve.

Natalie Simmons

Macaroni and Cheese Variations

#1 -
Brown 1/2 lb. ground beef. Drain off fat. Stir in one prepared macaroni and cheese mix and 1/2 cup taco salsa. Heat through.

#2 -
Stir together one 8 1/2 oz. can peas and carrots, drained; one prepared package macaroni and cheese mix; and one 6 1/2 oz. can tuna. Heat through.

#3 -
Stir together 1 prepared macaroni and cheese mix, 1/2 cup buttermilk salad dressing and 1 cup frozen peas. Chill in refrigerator for 3 - 24 hours. Stir in 1/4 cup milk to moisten, if necessary.

Cheese Sauce

1/2 C dehydrated cheese powder
3 T powdered milk
1 t. dried onions
1 C water
1 1/2 t. cornstarch
1 1/4 t. salt

Prepare as above recipe. Stirring until thickened.

Trading Post Pasta Skillet Dinner
(Dried Foods)

1 dehydrated beef patty or 1 C. ground beef gluten or 1 C. mock hamburger
1 1/2 C. elbow macaroni
2 T. dehydrated green peppers
1 C. tomato powder or 2 cans tomato sauce
1/4 C. dehydrated celery
2 T. dehydrated sweet corn
1/4 C. dehydrated onions
chili powder to taste
Italian seasoning to taste
salt to taste
pepper to taste

Reconstitute beef and vegetables. Boil macaroni until tender. Drain. Break beef into pieces. Sauté with onions, green pepper, and celery. Add tomato sauce, corn and seasonings and water as needed. Simmer 5 - 10 minutes. Add cooked macaroni and mix well.

One-Pot Spaghetti

1 lb. ground beef
1 T. dried onions
4 C. chicken bouillon
1 6 oz. can tomato paste
1/2 t. dried oregano, crushed
1/2 t. bottled minced garlic
6 oz. spaghetti, broken
grated parmesan or shredded cheddar cheese

Cook beef, drain off fat. Stir in chicken broth, tomato paste and spices. Bring to boil. Add spaghetti, stirring constantly. Reduce heat. Boil gently, uncovered for 15 - 17 minutes or until spaghetti is tender. Serve with cheese. Serves 4.

Tracie Bradley

Lasagne

1 Lb hamburger
1 onion, chopped
3 small can tomato sauce
1 tsp basil
1 tsp garlic salt
1 small box lasagne noddles
1/2 lb. Mozzarella
1 pt. cottage cheese

Brown onion and hamburger. Add tomato sauce, basil, and garlic simmer for 20 minutes. Cook noodles as directed on box. Layer noodles, sauce and cheese. Bake 350° for 45 minutes.

Noodles Alfredo

1/2 C. margarine
1/2 C sour cream
1 C. Parmesan cheese
1 T. parsley flakes
1/4 t. salt
1 sm pkg wide noodles

Cook noodles as directed. Heat margarine and sour cream over low heat until margarine is melted. Stir in remaining ingredients. Mix well, then gently stir in noodles. Serve.

Tomato-Mac

2 lb hamburger
1 onion, chopped
1 C. water
1 sm pkg. macaroni
2 cans tomato soup

Brown hamburger and onion: cook macaroni. Combine and add tomato soup and water; mix well. Heat and serve.

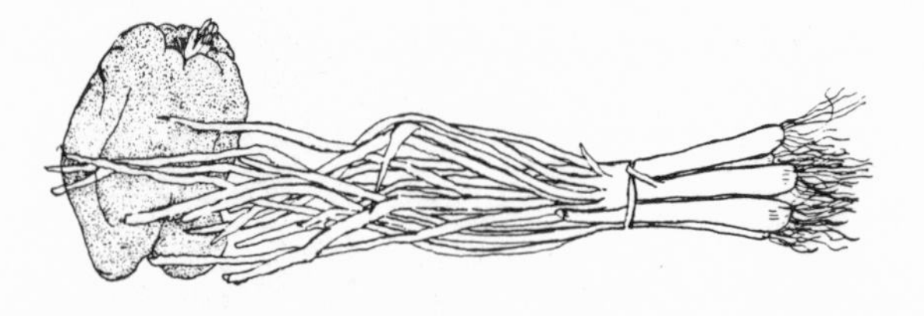

Soups were one of the most typical mainstays of the early settlers of Manti, the most famous being the Danish chicken and dumplings soup. Because supplies were scarce, the pioneers were very frugal. Any leftover vegetables, any soup bones and every scrap of meat was saved and made into soup. The soup was served with homemade bread, homemade butter, and bottled fruit or jam.

Chicken and Dumplings

(Pioneer recipe)

Soup

1 boiled chicken
2 onions diced
1 T. salt
1/4 C butter
1/4 C flour

Dumplings

1 C flour
1/2 t. salt
1 1/2 t. baking powder
1/2 cup milk
2 T melted butter

Cook the chicken, onion and salt until tender. Refrigerate over night. Skim the fat from the broth. (To speed the process, put a tray of ice cubes in the broth when it is hot, it will jell the fat and it can be skimmed off.) Remove the meat from the bones. In a separate pan melt butter, add 1/4 cup flour until thick, add 4 cups of the broth and cook until thick.

To make the dumplings, sift all dry ingredients together, add the milk and butter to form a dough. Drop the dumplings by spoonfuls into the cooked soup. Cook for 15 minutes or until the dumplings float to the top of the soup.

Peggy Layton

Danish Dumplings
(Pioneer Recipe)

1/2 square margarine (melted)
1 C flour
1 C hot broth or liquid
2 eggs

Stir all ingredients together and cook in a pan until thickened. Cool slightly and add 2 eggs, one at a time beating until a dough forms. Salt and pepper to taste. Drop by spoonfuls into hot soup. Do not boil. Dumplings are done when they float to the top of the soup.

Ruth Scow

Steamed Dumplings

1 C whole wheat flour
1/2 t salt
1 1/2 t baking powder
1 egg , beaten into 1/2 C milk

Mix all ingredients together. Drop by spoonfuls into boiling soup. Cover and steam 10 min.

Anna Jean Hedelius

Lumpy Dick
(Pioneer Recipe)

2 C milk (scalded)
1 T. margarine
1/4 t salt
1 dash pepper
1 egg , beaten
1/2 C flour
1/8 t . salt

Combine scalded milk with margarine, salt and pepper. In a separate bowl combine egg, flour, and salt together to form a dough. Break up into teaspoon size pieces and drop into the hot milk. Cover the pan and allow it to cook on low heat for 10 minutes. Serve in a bowl with cold milk or cream on it. Sugar and cinnamon can be added to taste.

Elta Alder

Cream Soup

2 T salad oil
2 T flour
2 T. butter or marg. powder
1/2 t. salt
3/4 C powdered milk
3 1/4 C water

Add the flour to the oil and blend. Add the butter or margarine powder. Real butter or margarine can be substituted. Stir in all remaining ingredients and cook on low heat until thickened. Serves 4.

Cream Soup Variations

Cream of celery soup: Add 1 C rehydrated celery and 1 T. minced onion. Fresh celery and onion can be used also.
Cream of mushroom soup: Add 1 C rehydrated mushrooms and 1 T. minced onion. Fresh mushrooms and onions may be used.
Cream of chicken Soup: Add 1 cup small pieces of cut up chicken and 1 T. minced onion.
Cream of potato Soup: Add 1 1/2 C diced cooked potatoes1 T. minced onion. Season with salt and pepper.

Cream of Onion (French Soup)

(Dried foods)

1 C dried onions
1 C boiling water
1 dash mace seasoning
1/4 C flour
1 quart milk
2 C water
2 chicken bouillon cubes
1 egg yolk, beaten
salt to taste

Saute onions in a little margarine. Sprinkle them with mace seasoning Stir in the flour, and cook over low heat until smooth and bubbly. Stir in the milk. Continue cooking until the mixture is slightly thickened. Stir in the 2 cups water with bouillon dissolved in it. Cook 5 more minutes and remove from the heat. Stir small amounts of soup mixed into egg yolk. (If you desire, 1 T egg powder mixed with 1 T water may be used). Season this to taste

Grated cheese can be placed in the bottom of the bowl before the soup is added. Top with dried bread crumbs or croutons.

Cream of Potato Soup
(Pioneer Recipe)

1 1/2 C cubed potatoes
1 T onion
3/4 C water
1 T. flour
1 T. margarine
3/4 t salt
2 C milk

Cook the potatoes, onion, and salt until tender. Blend the margarine and flour together and stir into the hot potato mixture. Stir constantly while cooking. When mixture thickens add milk and reheat. Modern day variation , (add beef or chicken bouillon to taste.)

PeggyLayton

Cream of Tomato Soup

2 C bottled tomatoes (no juice)
1/4 t. pepper
1 diced onion
2 t. brown sugar

White sauce
1 square margarine
1/2 C flour
2 C milk

Simmer all ingredients except white sauce for 15 minutes. In a separate pan make the white sauce by melting the margarine and adding the flour until a thick paste forms. Add the milk and continue to stir until thick. Pour the thick tomatoe mixture into the hot white sauce. Stir it thoroughly until mixed. Serve with hot homemade bread or crackers.

Tomato Soup
(Dried foods)

2 C milk
2 C dried tomatoes
2 T flour
1 T sugar
1 t. salt , dash of pepper
1 t. dried onions
1 dash garlic salt
2 T. butter

Heat the milk. Put dried tomatoes in a blender and add enough water to make 2 cups. Let this reconstitute for 5 minutes. Add remaining ingredients and blend until smooth. Add hot milk and blend.

Immigrant Cheese Soup

(Fresh or Dried)

1/2 C diced potatoes *
1/2 C diced carrots *
1/4 C celery *
1/4 C green onions*
1/2 T. butter
*** Dried foods can be substituted**

1/4 C flour
2 C milk
1 1/2 C chicken broth
1 1/2 C cheddar cheese *

If dried foods are used in this soup, they must be reconstituted first by adding warm water and letting them stand until they are soft.

Cook the vegetables in butter until tender. Pour flour into vegetable mixture and stir.. Cook until thick and bubbly. Add the cheese. Remove from heat. Stir until melted.

Marilyn Ostler

Mushroom Soup

(Dried Foods)

1 C dried mushrooms
2 quarts hot water
1/4 C dried onions
1/4 C margarine

4 cups dried diced potatoes
2 bouillon cubes
salt and pepper to taste

Pour hot water over mushrooms and onions. Let them stand for a few minutes to reconstitute. Simmer until the vegetables are tender. Drain the liquid and save it. Brown the mushrooms and onions in a small amount of margarine. Combine all the ingredients in a kettle and simmer for 40 minutes. Stir in remaining butter. Season to taste.

Peggy Layton

Cabbage Soup

(Pioneer recipe)

1 lb bacon or pork
1 cabbage
4 carrots
1 onion
6 potatoes
3 quarts of water
salt to taste

Thickening

1 pint milk
1/4 C bacon grease.
1 cup flour

Cut the pork or bacon into pieces and cook until tender. Wash cabbage and cut into chunks. Wash and cube the vegetables. Add to the water and meat and cook until tender. Combine milk, flour, and bacon grease together and add this to the soup to thicken it.

Plain Old Potato Soup

(Pioneer recipe)

2 quarts water
4 potatoes
2 carrots
2 onions
4-5 slices of bacon
2 T flour
2 C water
2 T. chopped parsley
salt and pepper to taste

Wash, peel , and cube the vegetables. Cook in the water until the potatoes are tender and mash in the water. Fry the bacon until crisp, add flour and cook until brown. Stir this mixture constantly so it doesn't burn. Slowly add 2 cups water and cook until thickened. Add this to the soup. Season to taste with salt, pepper, and parsley.

Peggy Layton

Most pioneer families had a root cellar which they filled with potatoes, carrots, onions, cabbage and any other vegetable that would keep through the winter. They also raised pigs, so bacon and pork were a main ingredient in many recipes.

Old Fashioned Vegetable Soup

(Dried Foods)

1/2 C dried onions
1/2 C dried celery
1/2 C dried green beans
1/4 C dried broccoli or zucchini
10 C water
2 dried beef patty (optional)
1 C dried carrots
2 C dried tomatoes
1/2 C dried corn
1 C dried cubed potatoes
3 t. beef bouillon
1/2 C rice or macaroni

Place all ingredients in a large kettle and simmer 10 minutes. 1/2 C. rice or macaroni can also be added. Add beef patties and bouillon. Simmer another 30 minutes or until vegetables are tender. Season to taste.

Vicki Tate

Ground Beef Soup

1/4 lb crumbled ground beef
1 onion, chopped
2 beef bouillon cubes
4 C canned tomatos
4 carrots, chop
4 potatoes diced
shredded cheese
salt and pepper to taste
1 t. basil
small bay leaf
1 1/2 C green beans
any left over vegetables

Brown the hamburger and onions. Season with salt and pepper. Cook the potatoes and carrots in 4 cups of salted water until almost tender. Add the rest of the ingredients except the cheese. Add green beans and onion. Simmer a few minutes . Add cheese to the top of each serving.

Peggy Layton

Poor Man's Soup
(Pioneer Recipe)

5-6 pieces of bacon
2 large potatos (diced)
2 stalks celery
1 lg. onion
1 quart water
salt and pepper to taste

Cook the bacon and break it into pieces. Pour off the grease. In a kettle, add bacon, and all remaining ingredients. Cook until tender.

Elta Alder

Danish Meatball Soup
(Pioneer Recipe)

Soup
6 carrots
4 potatoes
1 onion
2 stalks celery
1/2 t salt
water

2 C beef bouillon
2 T chopped parsley

Meatballs
1/2 lb hamburger
1/4 t sage
1/2 t salt
1/4 t pepper
1/2 slice bread
1 T cream or evaporated milk
1/2 T flour
1 egg

Wash, peel, and cut up vegetables. Cube the potatoes. Cook with 1/2 teaspoon of salt and enough water to cover until vegetables are tender.

Combine all ingredients for meatballs. Form into balls and fry until cooked.

Combine beef bouillon, parsley, meatballs, and vegetables together into a soup. Heat and serve this delicious meal.

Peggy Layton

Bean Chowder

3/4 C dry beans
3 C water
1 1/2 t. salt
3/4 C diced potatoes
1/2 C chopped onions
1 T. margarine
3/4 C bottled tomatoes
1/3 C green pepper
1 1/2 C milk
1 1/2 t. flour

Soak the beans overnight. Add salt and boil. Cover with a lid until almost done. About 1 hour. Add potato and onion. Cook 30 minutes more. Mix flour and margarine and stir into the beans. Add the tomatoes and green pepper. Cook over low heat about 10 more minutes until thickened. Stir in the milk and serve.

Marilyn Ostler

Clam Chowder
(fresh or dried foods)

1 C finely chopped onion *
1 C celery *
2 C diced potatoes *
2 cans clams
3/4 C butter
3/4 C flour
1 quart milk or half and half *
1 1/2 t. salt
pepper to taste
*** Dried or fresh foods can be used**

Place onion, celery, and potatoes in a pan. Drain the juice from the cans of the clams and pour it over the dried vegetables with enough water to cover them. Simmer them for 20 minutes or until done. In another pan melt the butter. Add flour, blend, and cook 1-2 minutes. Stir in the milk until smooth and thick. Add to undrained vegetables and clams. Heat again and add salt and pepper.

Vicki Tate

Sweet Corn Chowder

(Dried Foods)

1 1/4 C dried sweet corn
4 C water
2 t. dehydrated onions
2 T oil
2/3 C. powdered milk
2 T. bacon bits
1 t. salt
1/8 t. pepper
2 T. flour

Soak the corn and onion in water overnight. Add milk, salt, pepper, oil and bacon. Bring to a boil, reduce heat and simmer 30 minutes. Stir in the flour. Cook on low until it thickens, and the corn is tender. About another 15 minutes.

Jill Hanson

Salmon Chowder

1/2 C sliced onions
3 T. butter
2 C dried potatoes
1 C diced celery
1 can salmon
1/2 C diced carrots
3/4 t. salt
1/4 t. pepper
2 C water
4 C milk

Saute' onion in butter. Add potatoes, celery, carrots, salt and pepper to the water. Bring to a boil. Simmer for about 20 minutes with a lid on the pan until the vegetables are tender. Add the salmon with the liquid and the milk. This chowder can be thickened by adding equal amounts of butter and flour mixed together.

Wheat Chowder

2 C diced carrots
1 C boiling water
1/2 C diced salt pork
4 T. oni on
1 T. flour
2 C cooked wheat
1 t. salt
dash of pepper
2 C milk
1 T chopped parsley

Cook carrots in water unti l tender. Fry the pork until crisp and drain the fat. Keep 3 T. fat. Add onions and brown them lightly in the fat. Stir in the flour and thicken. Add all other ingredients and mix until well blended. Heat and serve.

Mary Fehlberg

Hopping Johns Bean Stew

1 C blackeye peas
5 C ham broth
1 C rice
1 C ham pieces

Soak beans in the broth for 1 hour. Cook until tender. Add rice and ham and cook 20-30- minutes. The broth should be almost gone when the rice is tender.

Marilyn Ostler

Mexican Wheat Chili Stew

1 lb ground beef
1 lg onion chopped
1/4 t. garlic salt
1 lg. can tomato sauce
4 T. flour
1 t. chili powder
1 t. salt
pinch of cayenne
1 t. oregano
1 t. Italian seasoning
2 C whole wheat soaked overnight
5 C beef bouillon , consumme, or mushroom soup
2 squirts A-1 sauce

Brown the hamburger. Drain and saute' onion with garlic salt. Add remaining ingredients. Simmer several hours until it reaches desired consistency. Add a can of green beans or corn 1 hour before serving. Check the stew often and stir. Adjust water if necessary.

Rebecca A. McGarry

Chow Wagon Beef Stew
(dried foods)

1 C beef flavored gluten or TVP or hamburger
2 C dried potatoes
2/3 C dried carrots
1/2 C dried peas
1 C dried onions
2/3 C beef bouillon
1/2 C flour
1/2 t. pepper
10 C water

Bring water to a boil in a large kettle. Add the vegetables, bouillon, flour, and pepper. Cook on medium until the vegetables are tender and the stew is smooth and thick. Add the meat or gluten the last 10 minutes of cooking time. The hamburger needs to be browned and crumbled.

Peggy Layton

Mom's Chili

2 lbs Red Beans
2 lbs hamburger
1 onion
salt to taste
1 T. chili powder
5 cans tomato sauce
2 C bottled tomatoes
pepper to taste

Cook beans until tender. Add salt and pepper. Fry hamburger with the onion. Drain the grease and add the meat to the beans. Add remaining ingredients.

Elaine Westmoreland

No Meat Chili

3 C Pinto beans
1 chopped onion
1/4 t. garlic powder
2 quarts of tomatoes
1 T. chili powder or seasoning
salt to taste

Sort and wash the beans. Simmer until soft with the onion and garlic. Salt the beans after they are tender. Add the tomatoes and chili powder and simmer for an hour longer. Serve with hot corn bread.

Betty Jenkins

Chili

2 1/2 C dried kidney beans
1 lb hamburger
1 C chopped onion
2 T chili powder
3/4 t. salt
1 T. Worcestershire sauce
3 C canned tomatoes

Soak the beans overnight. Cook until tender. Brown the meat and onions. Add seasonings and tomatoes. Cover and simmer for 10 minutes. Stir several times. Add beans and mix well.

Vicki Tate

Mormon gravy was a very important part of a pioneer supper. All fat and meat drippings were saved. If the fat was not used for gravy , it was saved for making soap. A large amount of gravy as made and put in the middle of the table at suppertime. Homemade bread was dunked into the gravy. It was used over meat, biscuits and potatoes, as well as any other vegetables available. Because pork and bacon were prevelant, the tasty fat and drippings made a base for the "Mormon Gravy ".

"Mormon Gravy"

(Pioneer Recipe)

8 T. Fat or meat dripping
6 T. flour
4 C milk
salt and pepper to taste

Fry the meat. Use an ample amount of fat and drippings, about 8 T. Add the flour and brown it slightly. Add the milk and stir until well blended. Season and cook to desired thickness.

Milk Gravy

(Pioneer Recipe)

1 C powdered milk mixed with 3 C water
3 heaping T. flour
1/2 t. salt
1 T. margarine

Mix the water and powdered milk together. Add the flour and salt.Cook over medium heat until the gravy is thickened. Add the margarine and stir until smooth

Ruth Scow

Basic Gravies

Chicken gravy

1 1/2 T. butter or margarine
1 1/2 T. chicken bouillon
1/2 t. garlic powder
1 1/4 C water
1/4 cup dried milk (instant)
1 t. onion powder
1/4 t. tumeric
1 T. flour

Onion gravy

1/4 C butter or margerine
1 T. beef bouillon
1 t. Kitchen bouquet (optional)
2 T. cornstarch
3 C chopped onions
2 C boiling water
1/4 C cold water
salt and pepper to taste

Beef Gravy

1 1/2 T. butter or margarine
1 t. minced onion or powder
1 t. Worchestershire sauce
1 1/4 C. water
1 1/2 T. beef bouillon
1/2 t. Kitchen bouquet
1 T. flour

Brown the onions if they are used. Add all other ingredients except cornstarch and cold water together in a pan. Cook until dissolved. Add the cornstarch and water together. Pour them into the gravy base. Continue cooking until thick and clear. Add salt and pepper to taste.

Peggy Layton

Bread and Gravy
(Pioneer Recipe)

Make gravy. Spoon over bread or corn bread.

In the early years when supplies were low, this was one of the dinners eaten most often. If available, a few vegetables were cooked and served along side.

"Mormon" Milk Toast

2 slices bread
1 t. flour
1 C. milk
salt

Cut bread in slices. Cut off the crust and toast it in the oven until brown. Butter the toast. Mix together 1 cup milk and 1 teaspoon flour and salt to taste. Cook till thick and pour hot over toast and serve.

Mormon gravy was eaten at almost every meal. It required few ingredients and was very versatile. The pioneers would serve it over bread, potatoes, biscuits or vegetables.

Pizza Sauce

1 8 oz. can tomato sauce
1/2 t. oregano
1/2 t. garlic powder

Blend all ingredients.

Lisa Conn

Basic White Sauce

Prepare Ahead Mix:

1 C flour
1 C margarine
4 t. salt
2 1/2 C nonfat dry milk

Blend ingredients with a fork until it resembles a crumbly coarse meal. Store in the refrigerator. To prepare : blend 1/3 C. mix with 1 C cold water or broth. Add liquid slowly. Heat to boiling, stirring constantly until thick.

Cathy Call

White Sauce

Thin	Medium	Thick
1 T butter	2 T. butter	3 T. butter
1 T flour	2 T. flour	3 T. flour
1 C milk	1 C milk	1 C milk
1/4 t salt	1/4 t. salt	1/4 t. salt

Over low heat, melt butter in a sauce pan. Add flour. Blend until smooth. Add milk at once and cook until thick. Stir constantly so it won't burn. Add salt. To make a cheese sauce add 1/2 C grated cheddar cheese.

Leanne Beal

Cheese Sauce

1/2 C dehydrated cheese powder
3 T powdered milk
1 t. dried onions
1 C water
1 1/2 t. cornstarch
1 1/4 t. salt

Prepare as above recipe. Stirring until thickened.

Pioneer Vegetable Sauce
(Pioneer Recipe)

1/2 lb bacon
2 T. onions
2 T. flour
2 T. sugar
1/8 t. salt
1/8 t. mustard

1 egg beaten
2 1/2 T. water
1/3 C vinegar
1 C evaporated milk
3 C cooked vegetables
or (pioneer greens)

Brown the bacon pieces with the onion. Combine all other ingredients together and add to the bacon and fat. Cook slowly until thickened and stir until smooth. Use this as a white sauce over the vegetables.

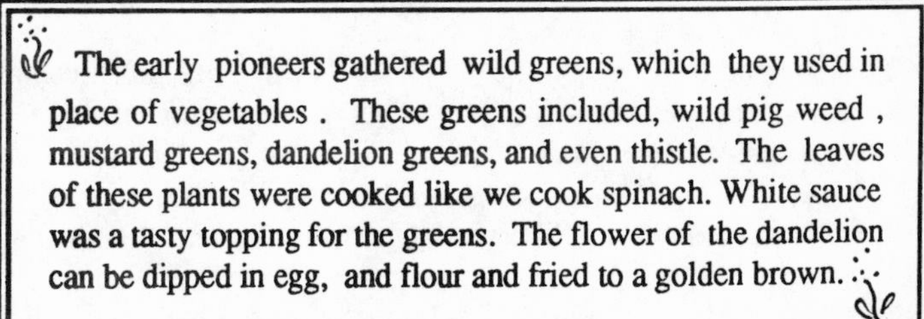

The early pioneers gathered wild greens, which they used in place of vegetables . These greens included, wild pig weed , mustard greens, dandelion greens, and even thistle. The leaves of these plants were cooked like we cook spinach. White sauce was a tasty topping for the greens. The flower of the dandelion can be dipped in egg, and flour and fried to a golden brown.

Pioneer Vitamin Sauce (for potatoes)
(Pioneer Recipe)

1 quart milk
4 slices bacon
5 T. flour

3 dandelion leaves
1 pinch sage

Put milk in a pan with the flour and sage. Wash the dandelion leaves thoroughly. When the sauce is cooked, take off the heat. Fry the bacon, break it into pieces and stir into the sauce. Cut dandelion leaves into 1/2 " pieces. Just before serving, stir the dandelion leaves into the sauce. Serve over potatoes.

Ruth Scow

Taco Sauce

(dried foods)

2 C tomato powder
4 C water
1/4 C minced onion
1/2 t. garlic powder
1 t. cumin
1 t. chili powder
1/4 t. nutmeg
2 T. vinegar
1 T. brown sugar
dash of salt
1 T. salad oil
1/4 t. pepper

Combine all ingredients and simmer until done.

Spaghetti Sauce

2 lbs ground beef
2 (16 oz) cans tomatoes
2 (6 oz) cans tomato paste
2 (4 oz.) cans mushrooms
3 T. seasoning mix

Brown ground beef and add all ingredients. Simmer 30 minutes.

Horseradish Barbecue Sauce

1 onion, chopped
1 C catsup
1 T. chopped fresh parsley
1 T. prepared horseradish
1 t. pepper
1 C water
1/2 C vinegar
1 T. brown sugar
1 T. prepared mustard

Barbecue Sauce

4 T worchestershire sauce
2 T mustard
1 C vinegar
1 C water
1/2 C sugar
1 C catsup
4 t. chili powder
2 t. accent
4 t. mi nced onion
1 t. Lawry's seasoned salt

Heat to dissolve all ingredients. Simmer 15 minutes. Use Horseradish sauce on pork or beef. Barbecue sauce on beans.

Pizza Sauce
(Dried Foods)

1 1/2 C. dehydrated tomato powder (reconstitued in 1 1/2 C. wàter)
1 t. salt
1 t. sugar
Dash pepper
1 t. Italian seasoning
1 bay leaf

Beat until smooth. A dd one bay leaf and refrigerate over night.

Sweet And Sour Sauce

1/2 C brown sugar
2 t. soy sauce
1 t. soup base or bouillon
2 T. cornstarch
1 C water
1/2 t. salt
3 T. vinegar
1 1/2 C pineapple juice

Mix brown sugar, soy sauce, soup base, and cornstarch in a saucepan. Add remaining ingredients and cook, stirring constantly until thick. Use this over pork or chicken for a wonderful Oriental dish.

Peggy Layton

Garden Fresh Chunky Salsa

Blend together until smooth and completely liquified:
Juice from 1 28 oz. can whole tomatoes (reserve tomatoes)
1-2 garlic buds
1 tsp. vinegar (optional)
1 dried chili pepper (Arboles or Japones depending on how hot)
Add and blend for just a few seconds:
1/3 bunch chopped fresh cilantro
1 (4 oz) can diced chiles
Add and blend on pulse or quick burst:
reserved canned tomatoes
Pour into a bowl and add to taste:
chopped green onions
diced tomatoes
diced green peppers

Mix well and store in a covered container in the refrigerator. Will keep for about 2 weeks.

Elaine's Chili Salsa

40 C. Tomatoes, peeled and cut up small
4 Large onions
18 cloves of garlic
5 long green peppers, with seeds
6 long banana yellow peppers
21 halapino peppers
8 C cider vinegar
4 1/2 T. salt

Run all ingredients except the tomatoes through the grinder and simmer for 5-7 hours.Process in a hot water bath for 25 minutes.

Elaine Harmston

Chili Salsa

2 T. oil
1 T. flour
1/2 C chopped onions
1 t. chopped garlic
3 1/2 T. chili powder
1/2 t. oregano
1/4 t. cumin
1/2 t. salt
1 C tomatoe puree
1 C beef broth

Heat the oil, add flour and cook 3 minutes. Stirring constaantly. Stir in the onion and garlic and cook until translucent. Add all other ingredients. Simmer 10 minutes.

Jill Hanson

Fresh Potato Chips

Big baking potatoes
Vegetable oil, peanut oil or melted shortening for frying
1/4 tsp. salt per potato, approximately

Wash and peel potatoes; cut them into paper-thin slices either all crosswise or all lengthwise for even cooking. To cut slices paper-thin, use your choice of a food processor, the slicing slot of a hand grater, a vegetable peeler, a very sharp knife, or a cabbage shredder/slicer. Rinse the sticky starch off the slices in a bowl under cold running water; pat them dry on paper towels.

Fry pan method: Pour equal amounts of oil and shortening into a frypan, heavy saucepan, or electric skillet to a depth of 3 inches; heat to 360°-375° on a frying thermometer. Fry potato slices, a handful at a time for 3-5 minutes or until golden and crisp; stir the slices after the spattering subsides to prevent sticking. Drain, salt, and repeat with the remaining slices.

Oven-baked method: Preheat oven to 450°. Very generously, grease 2 or more rimmed baking sheets with shortening; use enough to make a thick white film over each sheet about 1/8 inch deep; sprinkle lightly with salt to prevent spattering. Spread the potato slices, touching but in 1 layer, over the sheets; brush the tops of the slices with oil. Bake for 7-10 minutes or until golden and crisp. Drain on paper towels. Adjust salt to taste.

Variation After baking or frying chips, pour them immediately into a paper bag along with salt and your choice of: barbecue seasoning, chili powder, garlic or onion powder, or powdered cheese sauce mix. Shake the bag to coat, then drain on paper towels.

Au Gratin Potatoes with Sausage
(delicious)

10-12 potatoes
1 lb of sausage either Hillshire Farms round smoked or ground
1 large onion , chopped
1 can of cream of chicken soup
1 C sour cream
1 C evaperated milk
2 C grated cheese
1 t. salt or more to taste
pepper to taste
onion salt to taste

Boil the potatoes in the skins. When tender, cool them in cold water. Grate them into a cake pan or casserole dish. Mix all other ingredients together except onions and sausage. Mix only half the amount of cheese into the casserole. Spread the sauce over the potatoes and mix well. If you are using Hillshire Farms smoked sausage, slice the sausage into round pieces. If you are using ground sausage, brown and crumble it. Pour off some of the fat and fry the chopped onions. Add the onions and sausage last Fold it all in together and top with remaining cheddar cheese. Bake at 350° until cheese melts and casserole is bubbly.

Peggy Layton

Oven Baked Potatoes with Cheese
(Dried Foods)

3 C mashed potatoes or potato pearls , reconstituted
1/4 C dehydrated onion
3 1/2 T. dried cheese powder
1/4 t. paprika
1/2 t. salt
1/4 t. pepper
1/4 t. garlic powder

Soak the onions for 10 minutes. Add onion, spices , and 1/2 T. cheese to the potatoes. Mix well. Pour into a casserole dish. Rehydrate the remaining 3 T. cheese in 1 T. warm water. Spread over the potato mixture. Sprinkle with paprika. Bake at 350° for 15 minutes.

Elaine Westmoreland

Little Jake's Mashed Potato Boats

(Dried Foods)

4 C mashed potatoes or potato pearls
1/4 C dried butter or margarine powder
1/4 C dried cheese powder
2 T. dried onions (reconstituted)
Bacon or bacon bits (optional)
Salt and pepper to taste
Paprika

Shape 8 pieces of tin foil into boat shapes. To the mashed potatoes add rehydrated onions, margarine, cheese powder, salt, pepper, and bacon bits. Add hot water until the potato mixture is the desired consistancy. Spoon into the foil boats. Sprinkle paprika on them. Bake at 350 ° for 20 minutes until lightly browned.

Elaine Westmoreland

Old Fashioned Creamed Peas and Potatoes

(Dried Foods)

2 C cubed potatoes
1/2 cube butter or margarine
2 C milk
2 C peas
1/4 C flour
salt and pepper taste

Cut the potatoes in small pieces and boil until tender. Add peas. Cook a few more minutes. Drain the water from both. Put the butter or margarine in a sauce pan and melt . Add the flour to make a paste. Add the milk and stir until thick and smooth. Add salt and pepper to the peas and potatoes. Cheese can be added to the white sauce if desired.

Vicki Tate

Sausage Stuffed Baked Potato

Wash a bunch of baking potatoes. Rinse and dry. If a soft tender skin is desired, brush well with cooking oil. Bake in the oven at 450 until potatoes are tender. (About an hour). Cut baked potatoes in half lengthwise. Scoop out the insides and mash with butter, salt, pepper, and milk. Brown ground sausage. Drain the fat, and crumble. Mix the sausage into the potato mixture. Fill potato shells and place on a buttered cookie sheet. Toast under the broiler until browned. Just before done add a small amount of cheese to the tops of the potato boats. Melt and serve.

Connie's Candied Sweet Potatoes

1 large can sweet potatoes
3 T. flour
1 C evaporaated milk
marshmallows
1 C brown sugar
2 T. butter
1/2 C nuts

Mix sugar and flour well. Slice sweet potatoes in a casserole dish. Sprinkle sugar and flour over the potatoes. Dot with butter. Add nuts . Pour evaporated milk or cream over the top of it all. Top with marshmallows. Bake 40-50 minutes in 350 ° oven.

Connie Gardner

Golden Brown Hashbrowns

(Dried Foods)

1 3/4 C dried hashbrowns
1/4 C oil
2 1/2 C water
Salt and pepper to taste

Bring the potatoes and water to a boil for 10 minutes. Until most of the liquid is absorbed. Heat the oil in a skillet. Spread potatoes over the pan. Add salt and pepper to taste. Fry over low heat until potatoes are tender and golden brown.

Vicki Tate

Aunt May's Au Gratin Potatoes

(Dried Foods)

2 C dried potato slices
1 T. margarine powder
1/4 C powdered milk
salt and pepper to taste
2-3 C water
1 T. onion flakes
1/4 C cheese powder

Cook potatoes in water until tender. Saute onions in margarine until soft. Combine milk powder, cheese powder, and salt and pepper . Stir in enough water to make a sauce. Add to the potatoes and onions and mix well. Put in a casserole dish and bake until bubbly.

Vicki Tate

Fruits, Vegetables & Salads

Cooked Rhubarb

(Pioneer Recipe)

6 C rhubarb, cut into pieces
1 C sugar
water to cover

Cook the rhubarb until tender.

Rhubarb Mush

(Pioneer Recipe)

Fill a medium saucepan with cut rhubarb. Cover with water and cook until tender. Strain off the juice. For every 6 cups of juice add 1 cup of sugar. Place in a saucepan with 4 T. minute tapioca. Cook until tapioca is clear and juice is thick. When it is cool, serve with cream.

Ruth Scow

Fruit Leather

(Dried in the Sun)

Any bottled fruit

Drain the juice from the fruit. Put fruit in a blender and puree thoroughly. Line a cookie sheet with plastic wrap. Pour puree evenly over the plastic. (About 2 cups per cookie sheet). Set in the sun to dry or on a very low setting in the oven.

Cinnamon Apple Snacks

(Dried Foods)

dehydrated sliced apples
2 t. sugar or dry gelatin
1 t. cinnamon

Mix sugar and cinnamon in a bowl. Slightly wet the apple slices and dip them in the cinnamon sugar mix. Eat this for a fun snack. Another fun thing to do is to dip the apple slices in dry gelatin.

Peggy Layton

Lemon Honey Topping for Fruit

1/2 C whipping cream
3 T. honey
1/2 t. grated lemon peel
1 T. lemon juice

Combine whipped cream or topping with honey and lemon juice. Beat five minutes until soft peaks form. Fold in the lemon peel. Chill and serve over fresh fruit.

Christmas time was a joyous occasion for the pioneers. The early pioneer women began in the late summer and early fall picking wild berries, currents, and figs. They dried them to make sure they had enough for their Christmas puddings. Fruit trees and orchards were planted. Every home had several fruit trees on the property. Every pioneer home was equipped with a root cellar. The more hardy fruits were kept in the cellars through the winter. The more parishable fruits were dried to preserve them and were used in many tasty dishes. Apple pies were enjoy ed all year long because the apples kept so well.

Homemade Cinnamon, Applesauce
(Dried or Fresh)

2 lbs apples
1/2 C water or apple cider
honey to taste
lemon juice to taste
cinnamon to taste
*** Reconstituted dried apples may be substituted.**

Core the apples and cut them into chunks. Place apples, and water or cider in a saucepan. Simmer until tender. Force through an apple-sauce strainer or sieve. Season to taste with the lemon, honey, and cinnamon. The amount of sweetening will vary each time depending on the sweetness of the apples.

Vicki Tate

Chunky Applesauce

8 C peeled and cored apples
2 C water
2 t. lemon juice
1 t. cinnamon
honey to taste
1/2 t. cloves
1/4 t. nutmeg
1/4 t. allspice
1/4 t. mace
1/4 t. ginger

Put the apples, water and lemon juice in a kettle. Simmer until the apples are tender. Add remaining ingredients including enough honey to sweeten. Press the apple mixture with a potato masher, leaving some chunks for texture. Peaches and pears can be cooked in the same manner.

Lisa Conn

Reconstituting Dehydrated Fruits

1 C dried fruit
2 C water
2 T. sugar
1/2 t. lemon juice
*** For applesauce use 1 1/2 cups of fruit**

Add fruit to water in saucepan. Bring to a boil, reduce heat and simmer 10 minutes. Stir in sugar and cool, or cover fruit with water and let sit overnight in the refrigerator. Stir in sugar.

Everyday Fruit Cobbler

(Dried Foods)

Filling

2 C fruit
1 C sugar
2 T. butter
2 T. cornstarch
1/4 t. nutmeg

Batter

1 C flour
2 T sugar
1 1/2 t. baking powder
1/4 t. salt
1 egg
1 1/4 C milk
1/4 C shortening

Combine the filling ingredients and put in a square baking pan. Mix the batter ingredients together and drop by spoonfuls onto the fruit filling. Bake 20-30 minutes at 400 ° F.

Jill Hanson

Granola Apple or Peach Crisp

5 med. apples, sliced or 4 C canned peaches, drained
1/3 C flour
1/2 C brown sugar
1/3 C margarine, melted
1 t. cinnamon
1 1/2 C granola

Place apples or peaches in square baking dish. Combine other ingredients. Mix well. Sprinkle over fruit. Bake 25-30 minutes at 350 ° F. Serve warm or cold with milk or whipped topping.

Vicki Tate

Open Kettle Apples

(For Pie Filling)

12 quarts thick sliced apples
12 C sugar or less to taste
1 quart water

Cook all ingredients until translucent. Thicken with 1 cup of cornstarch in 2 cups of water. Can in the bottles while hot and bubbly, for 25 minutes.

For pies, pour in buttered bottom crust. This keeps the crust from getting soggy. Add cinnamon and nutmeg. Dot with butter. Wet the bottom rim before putting the top crust on.

Peggy Layton

Applets or Cotlets

2 1/2 C dried applesauce, apple nuggets, or apricots
5 t. unflavored gelatin
4 C sugar
2 t. vanilla
2 C chopped nuts
powdered sugar

Reconstitute the fruit. Cool. Soak the gelatin in 1 cup cooled fruit for 10 minutes. Combine the rest of the fruit with sugar and boil 10 minutes. Stirring often. Add gelatin mixture and boil another 15 minutes. Stirring constantly. Remove from heat and add nuts and vanilla. Pour into a greased, square pan. Set until firm. Dip in powdered sugar.

Jill Hanson

Fruit Cocktail Treat

(Dried Foods)

3 C. dried fruit mix
5 C. water
1 C. sugar
1/2 tsp. cinnamon
1/2 tsp. allspice
1/2 tsp. cloves

Combine all ingredients in a saucepan. Bring to a boil. Reduce heat and simmer 30 minutes. Serve warm with milk or whipped topping.

Elaine Harmston

Apple Snacks

(Dried Foods)

dehydrated sliced apples (or peach slices)
2 T. sugar
1 tsp. cinnamon

Mix sugar and cinnamon in a bowl. Moisten the apple slices and dip them in the cinnamon sugar mix. This makes a fun snack.

Variation Dip moistened fruit in flavored gelatin powder.

Dried Fruit Balls

1/2 C. dried peaches
1/3 C. dried apricots
2 T. graham cracker crumbs
1/2 C. whole, pitted dates
1/3 C. golden raisins

In a food processor bowl with metal blade, combine all ingredients except graham cracker crumbs; process until finely chopped. shape mixture into 1-inch balls; roll in graham cracker crumbs. Store in airtight container in refrigerator.

Cucumber Salad

(Pioneer Recipe)

2 large cucumbers
1/4 C water
1/2 t. salt
1/3 C cider vinegar
1/4 C sugar
dash pepper

Cut the cucumbers into slices. Mix all other ingredients together. Add the cucumbers and toss lightly.

Elaine Westmoreland

Quilters Potato Salad

(Pioneer Recipe)

Salad

6 potatoes
6 hard cooked eggs
1/2 C minced onions
salt and pepper to taste

Dressing

2 t. dry mustard
2 t. salt
6 T sugar
4 eggs
6 T melted butter
1 C hot vinegar
2 C heavy whipped cream

Salad: Cook potatoes whole. Cool, peel, and dice into small squares. Chop the eggs and add along with the salt, pepper, and onions.

Dressing: Combine all ingredients except whipped cream. Cook in a double boiler or on low heat just until thickened. Chill. Whip the cream until stiff. Mix with the dressing and the potato mixture.

Pioneer women often had get togethers with their women friends where they quilted for the day. At these "quilting bees" they were often served potato salad.

Fruit Salad
(Dried Foods)

4 C of reconstituted dehydrated fruits
4 C bottled fruit
2 C whipped topping

Mix fruits with either a whipped topping or a fruit sauce. Vanilla pudding is also very tasty over fruit.

Fruit Cocktail Treat
(Dried Foods)

3 C dried fruit mix
5 C water
1 C sugar
1/2 t. cinnamon
1/2 t. allspice
1/2 t. cloves

Combine all ingredients in a saucepan. Bring to a boil. Reduce heat and simmer 30 minutes. Serve warm with milk or whipped topping.

Tapioca Fruit Pudding
(Dried Foods)

1 C dried fruit mix
2/3 C sugar
1/8 t. salt
1/4 C quick tapioca
2 1/2 C water

Combine fruit, sugar, water, and salt in a saucepan. Simmer until fruit is tender, stirring often. Add tapioca and let it stand 5 minutes. Bring to a boil, stirring constantly for a few minutes. Take off the heat when pudding is thick.

Jill Hanson

Creamy Molded Pineapple Salad

1 pkg. lemon jello
2 C hot water
1/2 C mayonaise
1/2 C whipped cream
1 (# 2) can crushed pineapple
1 C grated cheddar cheese
1/2 C chopped nuts

Dissolve gelatin in the boiling water. Chill until partially set. Mix with mayonaise and whipped cream. In a greased mold, put a layer of gelatin, layer of pineapple, more gelatin, layer of cheese, more gelatin, and top with nuts. Serve with fruit dressing.

Fruit Dressing

Mix 1 1/3 T. flour, 1 1/3 T. sugar, and a pinch of mustard. Mix with one beaten egg. Add 1 C pineapple juice and heat to boil. Cool and fold in 1 C whipped cream.

Gelatin Fruit Salad

1 pkg. lemon gelatin
1 pkg. lime gelatin
2 C (pineapple juice and water)
1 C crushed pineapple, drained
1 C mayonaise
1 C cottage cheese
1 C evaporated milk
1 C chopped nuts

Heat liquid and dissolve jello in it. Allow jello to set, then whip jello and fold in remaining ingredients. Set in one large mold or several small molds. Allow to set.

Wonderful Fruit Salad Dressing

1 egg , beaten
1/3 C pineapple or pd. orange drink
1 1/2 T. lemon juice
1/2 C sugar
1/2 pint whipping cream
pinch of salt

Combine all ingredients and cook in a double boiler for 10 minutes or until it thickens. Cool and add whipped cream and salt. Add this to any gelatin or fruit salad as a topping or a dressing.

Peggy Layton

Cole Slaw

(dried foods)

Cole Slaw	**Dressing**
1 C dried cabbage	1/2 C mayonaise
3 C cold water	2 t. vinegar
1/2 t. celery seeds	2 t. sugar
1/2 t. sugar	1 t. salt

***Fresh cabbage may be used in place of the dried cabbage and water.**

Cover tightly and soak in the refrigerator one half hour. Drain if necessary. Serve with the cole slaw dressing mixed into the salad.

Fruit Salad

1 pkg small marshmallows
1 (#2 can) pineapple
2/3 can grapefruit sections
1 C grated cheese
2 whole eggs
3 T. vinegar
3 T. sugar
1 C whipped cream

***Orange sections and bananas may be added also.**

Add marshmallows, pineapple, and grapefruit. Combine with cheese. Beat egg yolks, vinegar, and sugar. Cook until thick. Cool and fold in the whipped cream. Combine this with the salad. Chill for one hour.

Crab and Pasta Salad

1 pkg curley noodles
1 small pkg immitation crab
4 green onions, chopped
4 stalks of celery, chopped fine
1 can olives, sliced
1 C frozen peas

Dressing

2 C ranch dressing mix
1/2 C milk

Cook the noodles. Cool them with cold water and drain. Chop the crab into chunks. Add all other ingredients together except dressing. Make the dressing and let it sit until thickened. Add this to the salad and serve.

Peggy Layton

Tomato and Onion Salad

Slice the tomatoes, sprinkle with chopped onions. Add a dash of salt, cider vinegar, and oil. Toss.

Cherie Harmon

Pinto Bean Combination Salad

6 oz lemon or lime jello
1/2 t. salt
1/4 C chopped green pepper
2 C cooked, soft pinto beans
3 T. vinegar
1 T. minced onion
1/4 C chopped pimento
1 sliced tomato

In a jello mold layer pimento, half of the pinto beans, chopped pepper and onion, remaining beans and tomatoes. Prepare jello and cool. Pour over beans and chill until firm. Serve over lettuce.

Foamy Green Salad

2 C. boiling water
1/2 C. sugar
1 can evaporated milk
1 C. crushed pineapple
1 (3 oz) pkg green jello
1/2 C. lemon juice
1/2 C. sugar
1 box vanilla wafers

Mix boiling water and jello together, add sugar and lemon juice, let set until cool. Beat 1 can evaporated milk until fluffy, add sugar, then whip in jello gradually. Fold in drained pineapple. Crush up vanilla wafers and save 1/2 C. Put the wafers in the bottom of a 9 x 13 pan. Pour jello mixture over wafers, sprinkle remaining crumbs over top and refrigerate until set.

Bean Salad

1 C cooked soy beans
1 C mung beans, sprouted
1 C chopped onions
1/2 C oil
garlic to taste
1 C pinto beans
1 C green beans
1 C chopped celery
1/2 C lemon juice
2 T. honey

Heat oil, lemon juice, honey, and garlic. Pour over beans and mix.

Mary Fehlberg

Ramen Noodle Salad

1 3 oz. pkg Chicken Oriental noodles
4 C shredded cabbage
3 T. vinegar
2 T. oil
1/2 C slivered almonds
4 green onions, sliced
2 T. sugar
1/4 t. salt

Crush the noodles slightly, place in a strainer. Pour boiling water over the noodles to soften them slightly. Drain well. In a large mixing bowl combine noodles, cabbage, and onions.

For the dressing: combine in a jar: The seasoning packet from the noodles, vinegar, oil, sugar, and salt. Shake well to mix. Pour over cabbage mixture and toss. Cover and chill several hours or overnight. Before serving, stir in almonds.

Tracie Bradley

Poppy Seed Dressing for Cabbage Salad

3/4 C sugar
1 t. salt
1 T. grated onion
1 t. poppy seeds
1 t. dry mustard
1/3 C vinegar
1 C oil

Mix dry ingredients together in a bowl. Blend with vinegar and onion. Add oil slowly as for mayonaise. Beating well after each addition. Stir in poppy seeds when mixture has formed a homogenous dressing.

Cathie Call

Delicious Dressing for Green Salad

2/3 C oil
1 t. salt
1 t. dry mustard
1/4 t. paprika
1/3 C cider vinegar
1 t. honey
1/2 t. basil leaf crumbled

Combine in a jar with a tight lid. Shake well and chill until needed.

Cherie Harmon

Thousand Island Dressing

1/2 quart Miracle Whip
1/2 C catsup
1/2 C relish
1 t. Worcestershire sauce

Mix all ingredients together and serve.

Creamy Salad Dressing

2 t. dried onion
1/2 t. salt
1 C buttermilk
1/8 t. garlic powder
1 T. parsley
1 C mayonaise

Put all ingredients in a jar and mix well. Store in the refrigerator.

French Dressing

1 can tomato soup
3/4 C vinegar
1 t. salt
1 T. Worchestershire sauce
1/2 t. pepper
1/2 C. sugar
1 T. prepared mustard
2 T. minced onion
1 T. paprika

Mix all ingredients well . Store in a jar. This will stay good in the refrigerator for 3 months.

Cathy Call

Note* Other dressing mixes are contained in the sauce section.

Danish Red Cabbage

(Pioneer Recipe)

1 head of cabbage, cut up
1 C sugar
1//2 C water
1/2 t. salt
1/2 C vinegar
2 T. shortening

Simmer until done. A little applesauce can be added for flavor while cooking.

Ruth Scow

Dandelion Greens

(Pioneer Recipe)

2 lbs. fresh dandelion greens
2 cloves chopped garlic
2 T. oil
salt and pepper to taste

The small young leaves are the most tender. Larger older leaves are bitter. Clean and wash the leaves. Do not eat the stem or the flower. Cut the leaves in half. Heat the oil and garlic in a saucepan. Add the leaves, salt, and pepper. Cook about 12 minutes or until tender. Add water if it gets too dry. Serve hot. A pioneer vegetable sauce can be put on top of the greens. See sauces for recipe.

Mary Felberg

Co-op Peas and Onions

(Dried Foods)

1 C dried peas
2 C water
salt and pepper to taste
2 T.. dried onions
1 T. margarine

Cook peas and onions in water until tender. Add other ingredients.

Green Tomatoes

Every year there are plenty of green tomatoes left on the vine when the cold weather comes. Here is one way to use them. Scald the tomatoes in boiling salt water. Peel and slice. Dip in flour and brown in a skillet with butter or margarine. Salt and pepper to taste.

Ruth Scow

Country Baked Carrots

(dried foods)

3/4 C dried carrots
1 1/2 T dried onion
1/4 t. salt
1/4 t. celery salt
2 T. flour
2 T dried cheese powder
2 C water
2 T. oil
dash of pepper
1/8 t. dry mustard
1 C milk
bread crumbs

Bring the water to a boil. Add carrots and simmer until nearly tender. (25 minutes). Drain . Rehydrate the onions and saute' them in the oil until tender. Add flour, spices, milk and cheese powder to the onions. Blend well and cook over medium heat until it boils. Pour into a greased casserole dish. Top with bread crumbs and bake uncovered until heated through. 15 minutes at 350 degrees.

Elaine Harmston

Aunt Susan's Simple Candied Carrots

(dried foods)

1/2 C dried carrot slices
6 T brown sugar
2 C water
1/2 C margarine

Cook carrots in water until tender. Add margarine and brown sugar. Heat until carrots are nicely coated.

Zucchini Creole

(Pioneer Recipe)

1 lb. zucchini squash
1 (# 2 can) tomatoes
3 slices bacon , crumbled
1 onion, chopped
salt and pepper

Peel zucchini squash and slice in 1/2 inch slices. Combine with tomatoes. Lightly brown the bacon pieces. Add chopped onion, zucchini, and tomatoes. Season and bake in a greased casserole dish at 350° F . for 45 minutes.

Zucchini Squash

(Pioneer Recipe)

1 lb zucchini squash
1 C med. white sauce
3/4 C grated cheddar cheese
1/2 t. Worcestershire sauce
1/4 t. mustard
salt and pepper

Scrub squash but do not peel. Slice into 1/2 " pieces and cook in salted water, until barely tender. Melt cheese into the white sauce over medium heat and add the seasonings to taste. Pour the sauce over the zucchini and serve.

Fried Eggplant

(Pioneer Recipe)

1 lg. eggplant
2 eggs, beaten
salt to taste
flour

Wash eggplant and slice paper thin. Soak in salt water for 15 minutes, then drain thoroughly. Dip into the beaten egg, then into the flour, and fry in hot oil until golden brown. Serve with sliced tomatoes.

Country Garden Vegetables
(Dried Foods)

3 C cabbage, fresh or dried
1 1/2 C diced carrots
1 1/2 C celery
1 T. honey
1/2 C onions
2 C water
1 t. salt
1/4 C oil

* Dried vegetables can be used. They need to be reconstituted first.

Combine vegetables. Add remaining ingredients. Cook gently for 15 minutes until tender. If using dehydrated vegetables add 2-3 t. sugar.

Cherie Harmon

French Fried Onion Rings

2 C cooking oil
3 large onions
1 C milk
1/2 C flour

Heat the oil slowly in a deep pan to about 375 degrees. Peel and slice onions 1/4 inch thick. Separate rings. Dip into the milk and then into the flour. Fry until golden brown. Drain well then place the onion rings on paper towels to absorb the excess oil. Salt and serve.

Summer Squash Casserole
(Pioneer Recipe)

1 lb summer squash
1 egg beaten
2 T. butter
salt and pepper
buttered cracker crumbs

Peel squash and cook until tender. Mash and mix with butter, beaten egg, salt, and pepper. Put mixture into a greased casserole and bake at 350° for 45 minutes.

Peggy Layton

Basic Sprouting Instructions

Day 1: Soak **1 T. of seeds or 1/3 C. beans in 1 qt. water** overnight.

Day 2: Rinse the seeds/beans thoroughly and drain. Place them in a quart jar and cover with a dampened cloth (cheesecloth works well). Fasten with a rubber band and store in a dark, cool place. Rinse the seeds/beans twice each day - three times if weather is very hot or dry. Make sure the excess moisture is drained off each rinse or the seed will ferment.

Day 3: The non-sprouters will sink to the jar's bottom during the rinse. Remove and discard them.

Approximate sprouting times:

wheat berries - 2 days
mung, lentil, soy, flax - 3 days
alfalfa - 4-5 days

When the sprouts are ready place them briefly in cold water and disentangle for use. For greener sprouts, expose them to sunlight for 2-3 hours before refrigerating. Store sprouts in covered container in refrigerator. Be sure to sterilize the jar before starting new seeds.

Nutritious Sprout Patties

2 C. wheat sprouts
1 egg, beaten
2 T. onion, minced
2 T. green pepper, minced
2 T. chopped mushrooms
oil
celery salt

Grind sprouts and add egg and vegetables. Mix well. Heat oil in skillet, form small patties and cook on each side to brown. Sprinkle with celery salt. Use as a main dish with a tomato sauce on them.

Wheat Sprout Meatballs

2 C. wheat sprouts
1 medium onion
1 t. salt
2 T. oil
2 eggs, beaten
2 C. bread crumbs

Grind bread crumbs. Put sprouts and onion through food grinder, using fine disc. Add salt, oil, and beaten eggs. Shape into balls and brown in oil in frying pan until brown and heated through.

Mary Fehlberg

Sprout Balls

1 C. ground nuts
1/2 C. sunflower sprouts
1/2 C. cream cheese
3 T. honey
1/2 tsp. vanilla

Mix, form into balls and chill. Can be rolled in toasted nuts, coconut, or granola. Makes 24 one inch balls.

Sprout Salad

alphalfa sprouts
bean sprouts
cubed tomatoes
sliced mushrooms
chopped green onions
sunflower seeds
bacon bits
lettuce or spinach
salt and pepper
avacado
2 hard boiled eggs

Combine and chill the following ingredients for dressing: 1/3 to 1/2 C safflower oil, the juice of 1 lemon, 2 1/2 T. honey, salt and pepper to taste. Toss the salad with the dressing and serve.

Sprouted Mung Bean Salad

mung bean sprouts
dried mushrooms

Reconstitute the mushrooms by covering them with water and heating until tender. Toss the mung beans and mushrooms in French dressing and marinate in the refrigerator for 1 hour.

Sprouted Mung Bean Salad

1 C. mung bean sprouts
1 C. fresh or frozen peas
1 C. coarsely grated carrots
1 C. chopped celery
1 large cucumber, diced
1 T. sesame seeds
2 C. buckwheat sprout "lettuce" or shredded head lettuce
Dressing - Fresh or oil and lemon

Toss lightly and top with dressing. Serves 4.

Rita Bingham

Super Easy Sunflower Salad

2 C. sunflower seeds (sprouted 2 to 3 days)
1 T. lemon juice
1 tsp. olive oil

Mix all ingredients (add salt to taste) and serve plain or on a bed of lettuce or sprouts.

Rita Bingham

Avocado and Sprout Sandwich

4 pita bread pockets
2 tomatoes, sliced
sliced cheddar cheese
2 avocados, peeled and sliced
2 C. alfalfa sprouts
mayonnaise

Open pita pockets all the way around and separate sides. Place cheese slices on one side and melt in oven. Spread other side with mayonnaise. Place avocado slices, tomato slices and sprouts on side with mayonnaise. Top with other side. Cut in half and serve.

Sauteed Sprouts

2 C sprouts
2 T. margarine
2 T. dried onions
2 t. soy sauce

Reconstitute the onions. Melt the butter and saute' onions and sprouts together. Stir in soy sauce. Serve.

Buttered Sprouts

1 C sprouts
2 T butter
1/2 C water
salt to taste

Simmer sprouts in salted water 3-5 minutes. Remove from heat and drain, add butter. Serve hot.

Sprouted Lentils

2 C canned tomatoes
1/2 C green pepper
1 t. molasses
1/2 C chopped onion
1 C tomato paste
2 C sprouted lentils
salt, pepper, paprika, celery salt, garlic.

Saute' green peppers and onions in 2 T margarine. Add remaining ingredients and simmer. Serve over rice or millet.

Dairy Products & Eggs
EGGS

Every pioneer family had a cow and chickens. These were brought with the first settlers to Manti. Generally the early settlers had plenty of milk, cream , butter, and eggs. Many of their meals included these dairy products. Sometimes on special occasions, chickens would be killed and cooked up into a wonderful soup or main dish.

Reconstituting Powdered Milk

For this amount	Mix dry milk		With water
1 quart	Inst. 1 C.	Reg. 3/4 C.	4 C.
1 pint	Inst. 1/2 C.	Reg. 1/3 C.	2 C.
1 cup	Inst. 1/4 C.	Reg. 3 T.	1 C.
1/2 cup	Inst. .2 T.	Reg. 1 1/2 T.	1/2 C.
1/4 cup	Inst. 1 T.	Reg. 3/4 T.	1/4 C.

Substitutions Using Powdered Milk

Whole milk

1 C water
1/3 C powdered milk

Evaporated milk and Whipped Topping

1 C water
2/3 C powdered milk

This milk can be chilled and whipped into a topping by adding 1/2 t. lemon juice. After it is whipped, fold in 1 T. sugar to taste.

Buttermilk

1 C water
1/3 C powdered milk
1 T. vinegar or lemon juice

Let mixture stand in a warm place until thickened (about 18 hours.) Stir until smooth. Refrigerate. A buttermilk freeze dried culture can be purchased at a grocery or health food store, and kept indefinately.

Soy Milk

Soy milk is a good substitute for any one that is allergic to cow's milk. Soak 2 cups of soy beans for 12 hours. Change the water frequently. Grind soaked, raw beans with a fine blade on a food grinder. Add 6 cups water to the beans in a large pan. Cook until foamy for 1 hour. Put through a blender, then strain through a cheese cloth. Refrigerate.

In the early days, before our modern refrigeration, food was kept cold in several ways. One cooling system was made by constructing a wooden frame about the size of a small refrigerator or ice box. The box was covered with wire and then covered with strips of burlap whose ends were placed in a bowl of water. The burlap on top of the box would act like a wick to keep the fabric damp around the box. As the breeze came through the windows and rustled the burlap, it created a nice cooling effect which kept the food cold.

Many of the older homes in Manti and around the area have a large rectangular slab of flat rock in their basements. This rock table stayed cool year round and was used by early settlers to keep their milk and other perishables cold.

In the neighboring town of Spring City, there are many natural springs. Some of the early settlers built their homes right over the top of them so they had running water in their basements at all times. They would channel the water into troughs to keep their milk cold. It also provided water for them year round without having to go fetch it. In case of Indian raids, they could hide in their basements for days at a time.

Sweet Sour Cream

1 C light cream **1 T. buttermilk**

Blend and put in a jar. Cover the jar and let stand for 24 hours at 80 °. Check the temperature often to make sure it stays constant. Refrigerate to set. Recipe can be doubled.

Sweetened Condensed Milk

1 1/3 instant dry milk or
3/4 C non- instant dry milk
3/4 C sugar
1/2 C hot water
4 T. margarine

Pour water into blender. Add milk and sugar. Blend. Add margarine and blend thoroughly. Chill for later use. 1 1/4 C homemade mixture equals 1 can regular sweetened condensed milk.

Debra Lindsay

Sweet Cream Butter

2 C whipping cream **1/4 C crushed ice**

Blend the cream in a blender or food processor or by shaking a jar full of cream until the cream separates from the whey. Pour off the liquid and discard. Start the blender and add ice until it forms a mass on the blades. Drain all the water and put the butter into a bowl and press to remove all liquid. Refrigerate for a day or two to age it. A little salt can be added to enhance the flavor.

Peggy Layton

Whipped Topping

1/2 C cold water
1/2 C powdered milk
1/2 C sugar
2 T. lemon juice

Put water in ice cold bowl. Add milk and beat with a cold egg beater until stiff. Add sugar slowly while beating. Add lemon juice and beat until soft peaks are formed.

Vicki Tate

Whipped Topping

6 T. powdered milk
2 t. gelatin (flavored or plain)
1/4 C sugar
1 C water
1 1/2 T. cold water
1 t. vanilla

Dissolve the milk and gelatin in boiling water. Add sugar. Stir and chill in the refrigerator until it jells. Beat the mixture until it looks like whipped cream. Add vanilla and whip again.

Elaine Harmston

A yogurt starter is a must. You can purchase a plain unflavored, freeze dried, yogurt culture from the supermarket or health food store. These starts can be used like a sourdough starter and fed every so often to keep them going indefinately. Keep the start stored in the frezer.

Yogurt

2 C. dry milk
2 T. plain yogurt
3 C. lukewarm water

Blend in blender and add 2 T. plain yogurt. Stir. Place in warm area (preferably 110 to 120°). Let stand undistrubed until set. The cooler the temperature the longer it takes to set. At 120° it can take about 4 hours. At room temperature it can take days. If it does not set it could be because: (1) It is not warm enough. (2) The yogurt culture used to make it was not fresh enough. (Seven to 10 days is getting pretty old.) (3) It was disturbed and so the structure was broken.

Yogurt from Non-Instant Milk

5 1/2 C water
1 12 C non instant milk **1 (12 oz) can evap. milk**
1 C plain unflavored yogurt **1/2 C gelatin (any flavor)**

Dissolve jello, 3 1/2 C hot water and powdered milk together until blended. Pour into a pan. Add remaining 2 C water. Add canned milk, heat to about 115 F. Remove from heat and add yogurt. Stir it in. Pour in 3 wide mouth jars . Put on the lids and set yogurt in a pan of warm water 125 covering to the height of the yogurt. Maintain water at above temperature for 4 hours. Refrigerate.

Debra Lindsay

Yogurt Ideas

Dip

1 C. Mayonnaise **1/2 C. yogurt**
1 T. Ranch style dressing mix

Makes a delicious dip.

Dressing

1C. mayonnaise **1 C. yogurt thinned with milk**
1 T. Ranch style dressing mix

Makes great dressing for a salad.

Popsicles

1 C. yogurt **1 can orange juice, undiluted**
1 C. fruit, pureed

Mix together. Pour into popcicle molds and freeze. Your kids will love it.

Flavored Yogurt

Yogurt may be substituted for sour cream in almost everything. Add a little onion salt and some dried parsley for use on baked potatoes. Use in stroganoff and cream tuna, etc. For flavored yogurt mix with a little jam or bottled fruit or flavoring or add about 1 t. jello (powdered) to 3/4 C. yogurt and stir. Kids love it!

Homemade Cottage Cheese

1/2 Junket Rennet tablet **1/2 C water**
1 gallon skim milk or reconstituted instant nonfat dry milk
1/2 C buttermilk **1 t. salt**
1/2 C cream (optional)

Disolve Rennet in water. Put the milk, buttermilk, and dissolved Rennet into large pan. Cover and let stand for 16-24 hours until curd forms and pulls away from side of the pan. Do not stir during this time. Cut curds with long knife into 1/2 inch pieces. Heat curd slowly over hot water until temperature reaches 110° F. (this takes about 1 hour). Hold curd at 110 ° F. until curd tightens. If it doesn't tighten or become firm, heat to 115 or 120 ° and hold at this temperature for 20 to 30 minutes. Stir every few minutes to keep curds uniformly heating. Pour mixture onto fine cheesecloth in a strainer and drain off whey. The whey can be used in breads or other foods where liquid is used. Dip the cheese cloth bag into cold water to rinse more of the whey off. Drain for 3 minutes. Place in a large bowl. Add salt and cream and mix thoroughly. Chill well.

Hints for making cottage cheese

Always use fresh milk. Reconstituted instant skim milk can be used. You must use a starter. Rennet is available in drug , grocery, or health food stores. (It is also called Junket tablets). Adding a small amount of cream to the cottage cheese makes a much better product. It's creamier and smoother and tastes better. Any pan used must not be made of aluminum or galvanized metal. Al l milk used must be pasteurized. Pasteurization will kill harmfüll bacteria. Most all dry milk that you buy has already been pasteurized.

To pasteurize whole milk. Skim off all cream from the milk. In a double boiler, heat the bottom water. Set the milk in the top boiler. Heat until it reaches 145 ° F. Keep the milk at this temperature for 30 minutes. Cool the milk to 72 ° F. by putting cold water into the bottom boiler. Use the milk immediately for best results.

If a sour acid taste remains, the curds were not rinsed sufficiently. Add a little salt to flavor it. If a yeasty, or bad flavor appears, it means that , yeasts, molds, or bacteria were introduced by using unclean utensils or the milk was not completely pasteurized.

Cottage Cheese
(Another Version)

2 C boiling water
1 C instant powdered milk
1/2 t. or 2,000 Mg. ascorbic acid powder (vitamin C)
1/3 C vegetable oil

Blend water and dry milk. Cook on medium heat. Add ascorbic acid and gently stir until milk curdles. About 20 seconds. Remove from heat and let rest 1 minute. Drain the whey off. The curds remaining are the soft cottage cheese. Moisten with buttermilk or cream and store in the refrigerator.

Peggy Layton

Quick and Easy Cheese

1 T. of plain yogurt per quart of pasteurized milk can be used as a starter bacteria. If you are using raw milk it will sour on its own. When the liquid separates and curds form, put it in a tight cloth bag and hang it so it will drip into a bowl. The curd that forms can be salted and used as cottage cheese or cream cheese. The liquid can be used in breads.

Shape the curd into a round ball and lay it on a cloth over a wire mesh so that the air circulates around the cheese. When it is firm, put it into a closed bowl in a cool place and leave it for 3-4 weeks. Turn it occasionally and taste it to see if it has aged. If mold grows on it, it is ok as it gives it flavor.

Medium Cheddar Cheese

6 C warm water
1 C vegatable oil
9 T. cheddar cheese powder
4 1/2 C dry milk
2 5/8 C white vinegar

Blend all ingredients except cheese powder. Pour into a hot greased saucepan and heat to 115 ° to form curds. Rinse the curds from the whey in warm water, then in cold. Add salt to taste and add the cheese powder and mix it well. Put into a cheese cloth and press it between two plates and a 1 pound object on top of the plate until all liquid is pressed out. Wrap in plastic and refrigerate.

Cream Cheese

1 quart whole milk
1/4 C buttermilk
1 T. water
2 C whipping cream
1 Rennet tablet

In a large saucepan combine milk, cream, and butermilk. Warm over low heat until mixture reaches 100 degrees. Remove from heat. Dissolve the Rennet tablet in the water. Add to the milk mixture and stir for 1 minute. Pour into a large glass bowl; cover and let rest undisturbed in a warm place for 16 to 24 hours or until firm curd has formed. A yellowish liquid will form around the curd.

Line a strainer with several layers of dampened cheesecloth large enough to be tied into a bag. Set the strainer over a bowl and drain the whey, then drain the soft curd until no more whey remains. Tie the corners of the cloth to make a bag. Hang it over the sink to drain. If the room is hot, drain over a bowl in the refrigerator.

Line the strainer with a clean piece of cheesecloth. Remove the cheese from the bag and mix the firmer outer cheese with the softer inner cheese. Place in the strainer and fold the ends of the cheesecloth over the cheese. Place a saucer on top and weight it down with an object weighing about a pound. Place a bowl under the strainer and refrigerate.

Let the cheese drain overnight or until it is desired firmness. Add more weight if needed. Remove the cheese from the cloth. Pack into a cheesecloth lined mold. Cover with plastic.

To unmold, lift cheese by the wrapping, peel back the cloth, and invert onto a dish. Store in the refrigerator. Eat within 3 days.

Peggy Layton

Zesty Parmesan Cheese

1 1/2 C boiling water
1 1/2 C dry milk
4 1/2 T. lemon juice

Blend together. Cook over low heat until milk boils. The curds will be small. Pour into a cheese cloth strainer. Rinse and pour out all excess water. Put curds in a bowl and stir to break up. Spread on a cookie sheet and dry for 120 minutes at 150 ° in the oven. Salt and blend in blender to make a powder. Put in a shaker.

Peggy Layton

Farmer Cheese

5 Rennet tablets
1 gallon whole milk (16 cups)
4 t. salt
2 T. warm water
1 small round can
4 C cold water

Rennet or Junket tablets can be purchased in a drug, grocery, or health food store. Dissolve Rennet tablets in 2 T. warm water. In a 5 quart Dutch Oven , heat the milk and Rennet mixture over med-high heat for 12 minutes or until the milk coagulates and separates from the whey and temperature reaches 120 . Stir often. Do not allow temperature to go above 125 . Pour curds and whey into a cheesecloth lined strainer. Drain well.. When the cheese is cool enough to handle , form into a ball, and squeeze out additional liquid. Put cheeseball in clean cheesecloth and place in a can to shape it . When it is completely cool, dissolve salt into water and place the salt water and cheese in a bowl. Refrigerate overnight. Drain, remove cheesecloth, and slice. This cheese will store 2 weeks in refrigerator.

Cheese Sauce

1 1/2 T. butter powder
1/2 C. powdered milk
1 1/2 C. water
1/2 C. powdered cheese
1 1/2 T. flour
1/4 t. salt
1/4 t. paprika

Mix all dry ingredients together except cheese powder. Add water gradually, stirring until blended. Bring to a boil and cook, stirring constantly, 1 to 2 minutes. Add powdered cheese and stir until smooth. Combine your favorite cooked vegetable with the sauce and pour over rice, or macaroni to make macaroni & cheese.

Buttery Herb Spread

1 C. softened butter
1/4 C. parmesan cheese
1/8 t. garlic powder
1 T. basil flakes
3 T. chopped chives
1/4 t. thyme

Mix all ingredients together and spread on sliced french bread. Wrap in foil , bake at 350° for 30 minutes. Then, if desired, toast in broiler until desired brownness.

In pioneer days, every family had chickens. Sometimes eggs were very abundant. The pioneers preserved them in several ways. To store eggs for an extended period of time, you must seal the pores of the shell.

Freezing is also an excellent way to keep eggs. Break them into ice cube trays and freeze. Store in a plastic bag.

Powdered eggs can be stored for a long period of time and used in all baking needs.

Reconstituting Powdered Eggs

Amount of eggs	egg powder	water
1 egg	2 T.	2 1/2 T.
2 eggs	4 T.	5 T.
3 eggs	6 T.	7 1/2 T.
4 eggs	8 T.	10 T.

Unflavored Gelatin

(A Substitute For Eggs)

To substitute for 1 egg : mix 1 t. gelatin, 3 T. cold water, 1/2 cup boiling water. For 2 eggs: mix 2 t. gelatin , 1/3 cup cold water, and 1/2 cup boiling water.

Before starting to mix cookies, cake, or something else : Place cold water in a mixing bowl and sprinkle gelatin in it to soften. Mix thoroughly. Add all the boiling water and stir until dissolved. While preparing the batter, place mixture in the freezer to thicken. When recipe calls for an egg , take the mixture and whip it until it is frothy. Then add it to the batter.

Whole Wheat Crepes

2 eggs
2 T. whole wheat flour
pinch of salt
1 C milk

Beat eggs and salt together. Add flour and beat slightly. Gradually add milk. Melt a small amount of margarine in a pan. Pour a thin layer of batter over entire surface of the pan. Cook over medium heat for a short time until bottom is slightly browned. Turn it over and cook lightly. Remove from the pan . Top crepes with a small amount of fresh fruit, powdered sugar, pudding, jam, tuna, chicken or even ice cream. Roll up and eat.

Nora Mickelson

French Omlet

3 eggs or (reconstituted egg mix)
1/4 t. salt
1 T. margarine
1 T. cold water
1/8 t. pepper
(optional) Chopped ham, sauted mushrooms, cooked onions, chopped tomatoes, and cheddar cheese.

Beat all ingredients except margarine. In a medium saucepan, melt the margarine. Pour in the eggs all at once. Let cook until the omlet is set . Add other ingredients such as chopped ham, sauted mushrooms, cooked onions, chopped green peppers, chopped tomatoes, and cheddar cheese. Fold the omlet in half with a spatula. Continue to cook over medium heat until cheese melts. Serve.

For a variation try 1/2 cup diced cooked potatoes. It makes a tasty potatoe omlet.

Scrambled Eggs
(dried egg mix)

8 T. egg powder
1 C water
4 T. dry milk
salt and pepper to taste

Stir all ingredients together. Melt 1 T. margarine in a pan . Scramble the eggs until done. Other ingredients that can be added are: Cheese, bacon bits, reconstituted peppers, mushrooms, onions or chives.

Desserts & Confections

Apple Fritters
(Dried Foods)

3 C flour
1/3 C dried eggs
1 1/2 C milk
4 t. baking powder
1 1/2 t. salt
2 t. sugar
2 T. oil
1/4 C water
1 C dried apple slices (reconstituted)

Sift together dry ingredients. Combine milk, oil, and water. Add to dry ingredients, mix well until moistened. Chop rehydrated apple slices and add to the batter. Mix. Heat 2 " of oil in a skillet and drop by spoonfuls into hot oil. Fry, turning once until brown. Drain on a paper towel. *Note* Reconstituted sweet corn can be added in place of apples to make corn fritters

Natalie Simmons

Apple Fritters
(Pioneer Recipe)

4 C flour
1 1/2 t. soda
1 t. salt
4 T sugar
1/2 t. nutmeg
4 eggs, beaten
2 2/3 C buttermilk
4 T. shortening
4 C apple chunks

Use above recipe instructions. Can substitute any fruit for apples.

Pioneer Raisin Sourcream Cheese Cake
(Pioneer Recipe)

1 Pastry crust
2 T cornstarch
3/4 C sugar
1/4 t. salt
1 t. cinnamon
1/2 t. nutmeg
1/4 t. ground cloves
3 egg yolks, beaten
1 C sour cream
1 C raisins
3 egg whites
6 T. sugar

Make a pastry shell for a 9 " pie and bake until lightly brown. In a saucepan, mix the first column of ingredients and the cloves together. Beat egg yolks and fold in. Stir in sour cream and raisins. Cook over low heat stirring constantly until thickened. Cool and pour into a pastry shell. Make a meringue out of the egg whites and sugar. Top with meringue. Brown in a 350 ° oven for 15 minutes until golden brown. Cool.

Peggy Layton

Turnover Delights
(Dried Foods)

1/2 C dried apples or peaches
1/2 C sugar
1 2/3 C water
2 T cornstarch
2 t. lemon juice
1/2 t. cinnamon
pastry shell

Boil water and apples in a saucepan for 2-3 minutes. Mix the rest of the ingredients except lemon juice and stir into apples. Boil 1 minute. Remove from stove and add lemon juice.

Roll out pastry to 1/8 " thick. Cut into 4" squares. Place a spoonful of filling in the middle of the pastry. Fold pastry over into a triangle. Moisten the edges and seal. Bake at 450 ° for 15 minutes or until lightly brown.

Natalie Simmons

Angel's Delight

(Dried Foods)

2 C dehydrated fruit
5 C water
1 small pkg. marshmallows
1/2 C sugar
1/2 C nuts

Bring water to a boil. Add fruit and cover, simmer for 30 minutes. Add the marshmallows, sugar, and nuts. Mix well and chill. Fold whipped topping into fruit mixture. Serve.

Natalie Simmons

Apple Kuchen

(Danish Dessert)

2 1/2 C flour
1 1/2 t. baking powder
1/2 t. salt
2/3 C sugar or honey
1 C margarine
1 1/2 C cooked oatmeal
2 beaten eggs
1/4 C milk

Custard

3 eggs
2 T. sugar
1/2 t. nutmeg or cinnamon
1 C milk

Stir together first four ingredients, cut the margarine into the flour to make a coarse crumb. Stir the cooked oatmeal into the mixture. Blend the eggs and milk to make a thick dough. Pat into dripper pan and along sides.

Arrange 2 1/2 cups of sliced apples on top of the dough. Sprinkle with a mixture of 1/2 t. cinnamon and 1/2 C brown sugar. Bake 350 ° for 10 minutes. Make the custard and pour over the top of the apples and continue baking until custard is firm.

Anna Jean Hedelius

Old Fashioned Apple Crisp
(Dried Foods)

4 C dried apple slices or nuggets
8 C water
1/2 C sugar
2 t. cinnamon
1/2 C dried marg. powder
3/4 C brown sugar
1/2 C oatmeal

Bring water, apples, sugar, and 1 t. cinnamon to a boil. Reduce heat and cover, simmer about 25 min. Drain off the liquid except 1 cup. Pour into a baking dish. Mix oatmeal, margarine powder, br. sugar and the other teaspoon of cinnamon. Sprinkle oatmeal mixture over the top and bake at 350 ° for 1 hour. You may substitute dried peaches or apricots for apples.

Elaine Westmoreland

Rhubarb Crunch

1 C rolled oats
3/4 C br. sugar, packed
1 t. cinnamon
1 C sugar
1 C water
1/2 C flour
1/2 C margarine
2 C diced raw rhubarb
2 T. cornstarch
1 t. vanilla

Mix rolled oats, flour, brown sugar, and cinnamon. Cut in the butter. Pat half the mixture onto the bottom of a greased baking pan. Cover with rhubarb. In a small pan cook sugar, cornstarch, and water. Stir in vanilla. Pour over the rhubarb. Top with the remaining crumbs. Bake at 350 ° for 45 minutes to 1 hour. Serve warm, plain, or with whipped cream or milk.

Leanne Beal

Lemon Pound Cake

2 C. sugar
1 C. butter
6 eggs
1 C. buttermilk
1 t. vanilla
1 lemon rind, grated
1/2 t. salt
1/2 t. soda
3 1/4 C. flour

Mix sugar, butter, and 3 eggs very well, add buttermilk, vanilla and remaining eggs. Add lemon rind, soda, salt, and flour. Mix well and pour into 2 well greased loaf pans. Bake at 350 ° for 1 hour.

Glaze

1 C. powdered sugar
Juice of 1 lemon

Mix well. While hot, poke holes in Lemon Pound Cake with a fork. Pour glaze over top.

Old Fashioned Gingerbread

1 C molasses
1 C shortening or oil
1 C sugar
1 egg
1 t. vanilla
2 T. soda in 1/2 C water
6 C flour
1/2 t. cloves
1 t. cinnamon
1/2 t. ginger

Simmer the molasses and oil together for 15 minutes. Cream together the next 3 ingredients. Sift all dry ingredients. Dissolve 2 t. soda in 1/2 C hot water. Add this to the sugar and egg mixture. Combine all ingredients. Stir until flour is mixed. Roll out and cut with a cookie cutter or shape for gingerbread cookies or house. Bake 10 minutes at 350 °. This recipe is good for children because the dough is easy to handle.

Judy Anderson

Topping For Gingerbread

2 T cornstarch
1 C orange juice or tang
1/4 C honey
1/2 C raisins

Blend cornstarch and honey. Heat with orange drink and raisins. Cook until thick. Serve hot over gingerbread.

Cherie Harmon

Mormon Johnny Cake

(Pioneer Recipe)

2 eggs, beaten
2 C buttermilk or sour milk
2 T. honey
2 T butter, melted
1/2 C flour
1 t. salt
2 C cornmeal

Beat eggs, add buttermilk , honey, and butter. Combine all dry ingredients together. Mix wet and dry ingredients together. Pour into a buttered baking dish or cake pan. Bake at 425 ° F for 20 minutes. Cut into squares and serve.

Very few ingredients were available to the early settlers. These cakes were simple and took the basic ingredients.

Peggy Layton

Coffee Cake

1/4 C shortening
1/3 C sugar
1 egg
2 C flour
2 1/2 t. baking powder
1/2 t. salt
2/3 C milk

Cream together first column of ingredients, add remaining ingredients to the creamed mixture. Stir well and pour into a greased pan. Sprinkle with spicy topping of 1/4 C sugar, and 2 t. cinnamon. Bake at 425 ° for 25 minutes. This cake can be baked in the microwave in 4-6 minutes. Turn the dish 1/4 turn every 1 1/2 to 2 minutes. Use grated orange rind instead of cinnamon for a nice variation.

Anna Jean Hedelius

Chiffon Cake

1 C egg whites (8 eggs)
1/2 C cream of tartar
8 egg yolks
2 C flour
1 C sugar
1 t. salt
3/4 C water
1/2 C cooking oil
1 t. vanilla
1 t. lemon rind

Beat egg whites and cream of tartar until stiff. Mix all other ingredients together in a bowl until smooth and shiny. Fold flour mixture into egg whites. Pour into ungreased 10 " tube cake pan. Bake 1 hour at 325 °. Increase temperature to 350 ° and bake 10 minutes longer. Turn cake pan upside down to cool.

Hint* Eggs in the chiffon cake need to be room temperature. They whip better and hold their shape better. Be sure the flour mixture is beaten shiny before folding into egg whites.

Anna Jean Hedilius

Sour Cream Cake

2 1/2 C. flour
1 t. soda
1/2 t. salt
1 C sugar
1 t. cinnamon
1/2 t. cloves
2 eggs
1 C buttermilk
1 C sour cream

Mix together and make a hole in the center. Then add wet ingredients. Beat until batter is smooth. Bake in 9" x 13" pan at 350 ° for 30 minutes. Use 2 T. chocolate instead of spices or add raisins or nuts for a variety.

Frosting

1/4 C margarine or butter
1/2 C packed br. sugar
3 T. cream or half and half
1/2 C chopped nuts
3/4 C flaked coconut

Mix together and spread on the cake warm. Place in the oven until it boils.

Anna Jean Hedelius

Any Old Bottled Fruit Cake

1 quart fruit blended with juice
2 C sugar
4 C flour
1 t. salt
1 t. nutmeg
nuts, rasins, dates, coconut
1 C oil
4 t. soda
4 t. cinnamon
1 t. cloves

Use fruit that has been sitting at room temperature. Sift dry ingredients and add to wet mixture. Bake in a greased and floured cake pan at 350 ° for 40 minutes. This is a good use for bottled fruit.

Leanne Beal

Any Fruit Upside Down Cake

1/2 C white flour
2 t. baking powder
1/2 C sugar
1/2 C milk
1/4 C melted shortening
1/2 C wheat flour
1/2 t. salt
1 egg
1 t. vanilla

Measure and sift dry ingredients. Combine egg, milk, vanilla, and shortening together. Turn into dry ingredients. Beat thoroughly and pour this over fruit sugar mixture in a baking pan. Bake 375 ° for 30 minutes. Serve with whipped topping.

Fruit Sugar mixture

4 T. butter
1 C brown sugar

suggested fruits: Apples, apricots, peaches, pineapple, or crushed dried fruits reconstituted. If apples are used sprinkle with cinnamon and sugar.

Vicki Tate

Whole Wheat Applesauce Cake

2 C wheat flour
1 C sugar
1 t. salt
1 t. cinnamon
1 t. soda
2 t. cloves
1/2 t. nutmeg
4 t. cocoa
1 C applesauce
1/2 C oil

Mix all ingredients together in a large bowl. Mix well and add apple sauce and oil and beat well. Put in an ungreased angel food pan or a 9x 13 dripper. Bake 35 minutes at 350 °. Use your favorite frosting or serve hot with whipped cream.

Nora Mickelson
Rebecca McGarry

Fudge Cake

2 C flour
6 T. cocoa powder
1/4 C dried egg powder mix
1 1/3 C water
1 1/2 C sugar
1 t. soda
dash of salt
3/4 C oil

Sift together dry ingredients, including egg powder. Add water and oil. Mix well. Bake in a greased cake pan for 40 minutes at 350 °.

Fudge Frosting

1 1/2 C sugar
1/2 C cocoa powder
1/4 t. salt
1 t. vanilla
1/4 C corn starch
2 T. margarine powder
2 C milk

Mix all ingredients, except vanilla, in a saucepan. Cook until thick, stirring constantly. Remove from heat. Cool slightly and add vanilla.

Vicki Tate

Sourdough Chocolate Cake

2/3 C shortening
1 2/3 C sugar
3 eggs
1 C sourdough start
3/4 C water
1 t. vanilla

1 3/4 C flour
2/3 C cocoa
1/2 t. baking powder
1 1/2 t. soda
1 t. salt

Cream together shortening and sugar. Add 1 egg at a time, beating well after each. Blend in the sourdough start. Sift together flour, cocoa, baking powder, soda and salt. Add alternately with water and vanilla. Mix at low speed of the mixer. Pour into a greased, floured pan. Bake at 350 ° for 35 minutes.

Elta Alder

Cocoa Buttermilk Cake

1/2 C butter or margarine
2 eggs
2/3 C baking cocoa
1 1/2 t. salt
1 1/2 C buttermilk

1 1/2 C sugar
1 2/3 C flour
1 1/2 t. soda
1 t. vanilla

Grease 9 x 13 cake dripper pan. Cream butter, and sugar until fluffy. Add eggs one at a time and beat well after each egg. In another bowl., combine flour, cocoa, soda, and salt. Mix well. Add dry ingredients alternately with the buttermilk and vanilla. Bake 20 to 25 minutes in a 350 ° oven. Cool in the pan and frost.

Nora Mickelson

Apples and Spice Pinto Bean Cake

2 C cooked Pinto Beans, mashed
1/2 C honey
1 egg
1/2 t. salt
1/2 C chopped nuts
1/2 t. each cinnamon, cloves, allspice
2 C applesauce
1/4 C oil
1 C whole wheat flour
2 t. vanilla
1 C rasins

Cream honey, egg, oil, and beans together. Mix with dry ingredients. Add applesauce, rasins, nuts, and vanilla. Bake as a cake or squares in jelly roll pan at 375 ° for 45 minutes.

Betty Jenkens

Shortcake

1/2 C sugar
3/4 C margarine
1 beaten egg
1 t. baking powder
2 T. corn starch
2 C sifted flour

Filling
2 C thick applesauce
1/2 t. cinnamon
1/2 C sugar
1 t. lemon juice

Cream together sugar and margarine. Add egg. Sift dry ingredients twice and add it to the creamed mixture. Divide dough in half. Roll each into a 9 " square. Place one in a 9 " cake pan. Mix the filling and put on top of the bottom crust and put the other crust on top as a cover. Bake at 350 ° for 30 minutes. Serve warm or cold. Excellent with ice cream or cream or plain.

Leanne Beal

Oatmeal Cake

2 1/2 C boiling water
2 C quick oats
1 C margarine
1 C brown sugar
1/2 C white sugar
4 eggs
2 t. vanilla
3 C flour
2 t. soda
1 t. salt
1 t. nutmeg
1 1/2 t. cinnamon

Mix the oats and water together and set aside for several minutes. Mix the next 5 ingredients together, add this to the oatmeal. Then add remaining ingredients. Bake on a cookie sheet for 30-40- minutes at 350 °.

Frosting for Oatmeal Cake

1/2 C margarine
6 T. evaporated milk
1 C brown sugar
1 C nuts
1 1/2 C coconut

Mix the first column of ingredients until butter is melted. Then add nuts and coconut. Mix all together and spread over hot cake. Place back in the oven for 2 minutes.

Louise Eddy

Carrot Cake

1 C butter, softened
1 t. cinnamon
1 t. mace
3 t. baking powder
1/2 C honey
3 eggs
2 C whole wheat flour
1/2 C hot water
2/3 C chopped nuts
1 lg can crushed pineapple
1 C grated carrots
1 C raisins, soaked in water

Cream Cheese Icing

1/2 C soft butter
1/4 C + 2 T honey
1 t. vanilla
8 oz. cream cheese, softened

Soak raisins in 1 C hot water. Drain off water . Beat butter, spices, and baking powder together. Beat in honey gradually. Beat in eggs one at a time. Stir in the 1/2 cups hot water and flour. Add nuts, pineapple, carrots, and raisins. Bake at 350 ° for 35-45- minutes in a greased cake pan. Beat icing ingredients together. Frost the cake.

Debbie Wade

Yellow Cake

2 C flour
3 t. baking powder
1/2 C oil
1 1/2 C water
1 1/2 C sugar or honey
1 t. salt
1 t vanilla
2 eggs

Sift together dry ingredients. Add shortening, oil, and eggs. Mix well. Pour into a greased cake pan and bake at 375° for 25-30 minutes.

Frosting

4 C powdered sugar
4 T. milk
2/3 C marg. powder
1 t. vanilla

Rehydrate the butter or margarine powder with 2 t. warm water. Mix well. Add powdered sugar, milk, and vanilla. Beat until fluffy and smooth. Add more milk as needed. Add cocoa powder for chocolate frosting.

Elaine Harmston

Apple Coffee Cake

(dried foods)

3 C flour
2 T. baking powder
1/3 C dried egg powder
1 1/2 C water
1 1/2 C sugar
1/2 C shortening
1 1/2 T. salt
1 C dried apple slices reconstituted and chopped

Blend all ingredients together and mix well. Pour into a greased baking dish. Sprinkle the top with topping mix. Bake at 375 ° for 25-30 minutes.

Topping Mix

2/3 C brown sugar
3/4 C mar. powder
2 T. water
1/2 C flour
1 t. cinnamon

Elaine Harmston

Poor Man's Cake

(Old Nauvoo Recipe)

2 C. water
2 C. sugar
2 C. raisins
3 T. lard
1 tsp. cinnamon
1 tsp. nutmeg
1 C. nuts
3 C. flour
1 tsp. baking soda
1 tsp. salt

Boil first four ingredients together. Let cool. Then add the dry ingredients that have been sifted together. Add the nuts. Bake 1 hour at 350° .

Sugared Bread Cake

1/2 C. warm water
1/2 t. sugar
2 T. dry yeast
3/4 C. warm water
1/2 C. sugar
2 T. dry milk
1/4 C. dry instant potatoes
1 t. salt
2 eggs
2 3/4 C. flour
1/2 C. butter
3/4 C. brown sugar
1 t. cinnamon

Take first 3 ingredients and mix together and set aside until yeast bubbles. Mix brown sugar and cinnamon together and set aside. Melt the butter and set to let cool. Take water, sugar, dry milk, dry potatoes, salt , eggs , yeast mixture, and 1 C. flour and beat 2 minutes. Add remanining flour, mix well. Place in greased bowl, grease top of dough, cover and let rise until double, about 1 hour. Punch down and put in greased shallow pan (17 x 12 -1 " deep). Let rise 30 minutes. Spread evenly in pan and sprinkle evenly with sugar mixture. Make shallow indentations with fingers and dribble with butter. Let rise 30 minutes and bake at 375° for 12-15 minutes, until golden brown.

Whole Wheat Pineapple Upside Down Cake

2 C. sugar or 1 C. honey
2/3 C. butter
2 eggs
1 1/2 tsp. vanilla
1 (20 oz.) can sliced pineapple (drain, reserving juice)
3 C. whole wheat flour
1 tsp. salt
3 tsp. baking powder
2 C. milk

Cream sugar, butter, eggs, and vanilla. Add dry ingredients and milk, mix. Place pineapple slices in a 9x13 pan with 3 T. juice. Pour batter over fruit. Bake at 350° . for 30-40 minutes.

Whole Wheat Fruit Shortcake

2 C wheat flour
1/2 t. salt
1 egg
1 T brown sugar or honey
3 t. baking powder
1/3 C margarine
1/2 C milk

Mix together the first three ingredients. Cut in the margarine until it resembles coarse crumbs. Beat egg, milk, and sugar or honey . Add to the dry ingredients and stir until moistened. Roll on a floured surface and cut as for biscuits. Spread tops with butter. Bake at 425 ° for 15-20 minutes.

While it is hot, serve with strawberries, canned fruit or reconstituted dried fruit. Top with whipped topping.

Elaine Westmoreland

"Yummy to the Tummy" Chocolate Chip Cookies

2 C brown sugar
2 C white sugar
2 C butter
4 eggs
2 t. vanilla
1 t. soda
4 C whole wheat flour
2 t. baking powder
5 C oatmeal, Put in a blender and grind into flour.
1 C walnuts
2 C chocolate chips

Cream butter, sugars, eggs, and vanilla together. Add dry ingredients and stir until stiff. Add walnuts, and chocolate chips. Use a teaspoon to form balls on a cookie sheet. Bake at 400 ° for 8 minutes. Take off cookie sheet immediately and let stand until cool.

Marianne Thompson

Brown Sugar Banana Cookies

(dried foods)

1 C dried bananas
1 C boiling water
1 3/4 C wheat flour
2 t. baking powder
1 1/2 C oatmeal
1/2 C margarine
1 C brown sugar
1/2 t. salt
1/2 t. nutmeg
1 egg, beaten
1 t. vanilla

Pour boiling water over bananas and let stand for 10-15 minutes. Sift together all dry ingredients. Cream the butter, sugar, egg, and vanilla and add to the other ingredients. Add oatmeal and mix well. Drop by teaspoon onto a cookie sheet. Bake at 350 ° 12-15 minutes.

Natalie Simmons

Ground Raisin Cookie

4 C rasins
1 1/3 C water
2 C sugar
1 C oil
4 eggs
2 t. vanilla
4 C flour
2 t. cinnamon
2 t. nutmeg
1 t. salt
2 t. soda
1 t. baking powder
2 C quick oatmeal

Boil the raisins and water for 2 minutes. Strain the raisins and grind. Place them back in the water. Cream sugar, oil, eggs, and vanilla together. Add the raisins and water to creamed mixture. Add remaining ingredients. Mix well.. Drop by spoonfuls on cookie sheet. Bake at 375 ° for 10 minutes.

Louise Eddy

Elaine's Sugar Cookies

(dried foods)

2 1/2 C flour
1 t. baking soda
1 t. cream of tartar
1/4 t. salt
1/3 C water
2 C margarine powder
1 1/2 C powdered sugar
3 T dried egg mix
1 t vanilla
1/2 t. almond extract

Cream together margarine, sugar, and eggs. Add water, vanilla, and almond extract. Beat until creamy. Sift all dry ingredients together. Add to creamed mixture and mix well. Place on a lightly floured surface and roll out 1/4 " thick. Cut into desired shapes. Bake 7-8 min. at 375 °. Frost if desired or sprinkle with sugar.

Elaine Harmston

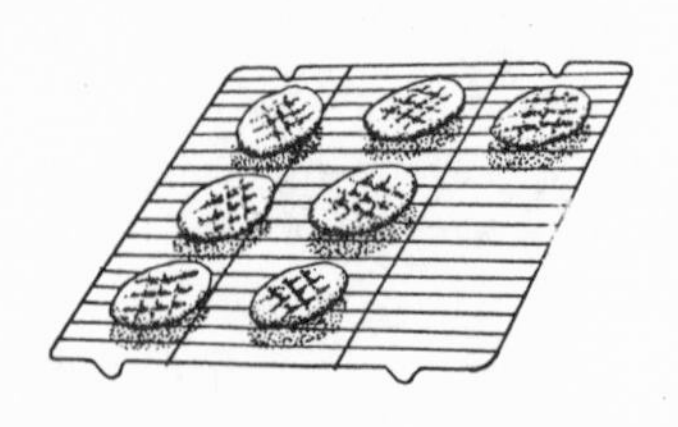

Dee's Favorite Oatmeal Cookies

1 C br. sugar
1 C white sugar
1 t. vanilla
3 C rolled oats
2 t. cinnamon
1 3/4 C. flour
1 C. shortening
1/2 C chopped nuts (opt.)
1 t. salt
1 t. soda
2 eggs

Beat sugars and shortening until well blended. Add remaining ingredients and mix well. Roll into balls and smash or roll into rolls, refrigerate and cut into slices onto a cookie sheet. Bake at 400 ° for 9 minutes.

Elaine Harmston

Raisin Oatmeal Cookies

1/2 C shortening
3/4 c sugar
2 eggs
1 1/4 C milk
1/2 t. salt
1 C oatmeal
2 C flour
1 t. soda
1 t. cinnamon
1 C raisins

Cream together the first column of ingredients. Add the remaining ingredients. Beat thoroughly and drop onto a greased cookie sheet. Bake 12 minutes at 350 °. You can substitute applesauce or any other blended fruit in place of the milk, or add chocolate chips.

Anna Jean Hedelius

Apple Oatmeal Cookies

1/2 C butter
1 C br. sugar
2 eggs
1 3/4 C wheat flour
1/2 C rolled oats
1/2 t. salt
2 t. baking powder
1/2 t. cinnamon
1 C raisins
1 C chopped walnuts
1 1/2 C chopped apples

Cream butter, sugar, and eggs. Combine all remaining ingredients and stir . Drop by teaspoons onto a cookie sheet and bake 12-15 minutes at 350 degrees.

Vicki Tate

Peanut Butter Oatmeal Cookies

Master Mix on page 74

2 C oatmeal cookie mix
1/2 C peanut butter
1/4 C honey
1/2 t. vanilla

Mix all ingredients together. Drop by spoonfuls on a cookie sheet. Bake at 375 ° for 10-15 minutes.

Elaine Harmston

Oatmeal Chocolate Chip Cookies

Master Mix on page 74

2 C oatmeal cookie mix
1/4 C evaporated milk
1 C chocolate milk

Combine ingredients and mix well. Drop by spoonfulls onto a cookie sheet. Bake at 375 ° for 10-12 minutes.

Banana Coconut Cookies

Master Mix on page 74

2 C basic cookie mix
2 T. dried egg mix , reconstituted
1/2 C coconut
1 t. vanilla
1/2 C chopped nuts, or rolled 6 grain cereal
1/4 C dried crushed banana chips, softened in 2 T water

Mix all ingredients together. Drop by spoonfuls on a cookie sheet. Bake in 375 ° oven for 10-12 minutes.

Natalie Simmons

Ginger Snaps

1/2 C shortening
3/4 C sugar
1 egg
13 C molasses
2 1/2 C flour
1/4 t. salt
1 t. cinnamon and ginger
1/2 T. cloves
2 t. soda
1 T. water

Mix all ingredients together. Pinch off the dough the size of a walnut. Roll in the sugar. Bake on the cookie sheet at 350 ° for 10 minutes.

Anna Jean Hedelius

Peanut Butter Cookies

1 C softened butter
1 C peanut butter
1 1/4 C honey
2 eggs
1 t. vanilla
4 C whole wheat flour
1/2 t. salt
2 t. soda

Blend butter until smooth. Add peanut butter and blend again. Add honey and blend again. Add eggs 1 at a time and beat until mixed. Add vanilla. Add dry ingredients and mix well. Roll into balls and put on ungreased cookie sheet. Use a fork dipped in cold water and flatten with a criss cross of the fork. Bake at 350 ° for 10 minutes on the top rack of the oven. Remove immediately. Cool and enjoy.

Debbie Wade

Peanut Butter Cookies

(dried foods)

1 1/2 C peanut butter or reconstituted peanut butter powder
1/2 C sugar
1/3 C shortening
1 1/4 C flour
1/2 t. baking powder
1/2 t. salt
1/2 C brown sugar
3 T. dried eggs
2 1/2 t. soda
1 t. ginger
1/2 C water

Mix Peanut butter, shortening, sugars, egg, and water in a bowl until creamed. Sift dry ingredients and add to the creamed mixture. Mix well. Roll into 1 1/2 " balls and flatten with a fork in a criss cross pattern. Bake at 375 ° for 10-12 minutes

Natalie Simmons

Snickerdoodles

1 C shortening
1 1/2 C sugar
2 eggs
2 3/4 C flour
1 t. soda
2 t. cream of tarter
1/4 t. salt
2 T. sugar
1 t. cinnamon

Mix shortening, sugar, and eggs together. Sift dry ingredients and mix with creamed mixture. Form into balls the size of walnuts. Mix 2 T sugar with the cinnamon. Roll balls in mixture and bake on an ungreased cookie sheet at 400 ° for 10 minutes. Don't overbake.

Mary Lois Madson

Sourdough Spice Nut Cookies

1/2 C sourdough start
1/2 C brown sugar
1/4 C shortening
1/2 C molasses
1 egg
2 cups flour
1/2 t. salt
1 t. ginger
1/2 t. cinnamon
1/2 t. nutmeg
1/2 t. cloves
1 t. soda

Cream together first 4 ingredients. Add all other ingredients and 1/2 cup of nuts . Bake on a cookie sheet at 375 ° for 1 minute. These cookies may be frosted.

Anna Jean Hedelius

Pumpkin Cookies

1 1/2 C brown sugar
1/2 C. shortening
2 eggs
1 3/4 C pumpkin
1 t. baking powder
chocolate chips or nuts
1 t. cinnamon
1/2 t. nutmeg
1/2 t. salt
1/4 t. ginger
2 3/4 C flour

Cream first 4 ingredients. Add remaining ingredients. Bake at 400 ° for 12 minutes.

Louise Eddy

"Debbie's Favorite"

Whole Wheat Oatmeal Cookies

2 eggs
1 t. vanilla
1 C butter
3/4 C honey
1/2 C pure maple syrup or honey
1/2 C fructose

3 1/2 C whole wheat flour
1/2 t. salt
1 t. soda
1 t. baking powder
4 1/2 C rolled oats
1/2 C carob chips
1/2 C chopped nuts

Cream all wet ingredients together. Add to the dry ingredients and blend well. Add carob chips and nuts last. Stir. If the dough is too sticky, add 1/2 C more flour and 1/2 C more oats. Drop by spoonful onto an ungreased cookie sheet and bake at 350 ° for 15 minutes. Do not overcook.

Debbie Wade

Honey Cookies

(Dried Foods)

1 1/3 C margarine
1 C honey
2 eggs
1/2 C milk
3 1/2 C white flour
1/4 t. salt
1 t cinnamon
1/2 t. allspice

2 t. baking powder
1/4 t. salt
1/2 t. soda

1 C chopped mixed dried fruit such as: raisins, dates, apricots, or apples, reconstituted. The crumbs from fruit mix will work.

Combine butter, and honey. Stir in the eggs and milk. Sift together dry ingredients. Add the fruits to the flour mixture and stir into the honey mixture. Drop by teaspoons onto a cookie sheet. Bake at 375 ° for 12-15 minutes.

Vicki Tate

Ginger Cookies

Master Mix on page 74

2 C basic cookie mix
1/2 t. ginger
1/2 t. allspice
1 egg, beaten or 2 T. dried egg mix
1/4 C molasses
1/2 t. cinnamon
1/2 t. vanilla

Combine all ingredients and mix until blended. Drop by spoonfuls on the cookie sheet. Flatten with bottom of a glass dipped in sugar. Bake at 375 ° for 8-10 minutes.

Natalie Simmons

Chocolate Raisin Cookies

Master Mix on page 74

2 C basic cookie mix
1/4 C sugar
2 1/2 T water
1/4 C baking cocoa
1/2 C raisins
1 t. vanilla

Combine all ingredients and mix until moistened. Drop by spoonfuls onto a cookie sheet and bake for 10-12 minutes at 350 °.

Natalie Simmons

Cowboy's Chocolate Chip Cookies

(Dried Foods)

1 C whole wheat flour
1 C white flour
1 C shortening
3/4 C white sugar
3/4 C brown sugar
1/4 C dried egg mix
1/4 C water
1 t. baking soda
1 t. salt
1 t. vanilla
1 C oatmeal
1 C chocolate chips

Cream shortening and sugars together. Add water and all other ingredients. Add oatmeal and chocolate chips last. Bake at 375 ° 10-12 minutes.

Jill Hanson

Gumdrop Chews

2 C oatmeal Cookie Mix
1/2 C diced Gumdrops
2 T dried egg mix + 2 1/2 t. water or 1 egg, beaten
1/2 C coconut
2 T milk

Mix all ingredients together. Drop by spoonfuls onto a cookie sheet. Bake at 375 ° for 10-12 minutes.

Elaine Harmston

No Bake Cookies

3 T. cocoa
1/4 C. margarine
1/4 C. milk
3/4 C. sugar
1 1/2 C. quick oats
1/2 C. coconut

Cook cocoa, margarine, milk, and sugar over medium heat, stirring occasionally until mixture comes to a full boil. Remove from heat and add oats and coconut, drop quickly by spoonful onto wax paper. Let cool.

Oatmeal Honey Cookies

1 C. shortening
1 1/2 C. honey
2 eggs
1 t. cinnamon
1 1/2 C. flour
1/2 t. nutmeg
1 t. baking powder
1/2 t. salt
3 C. oats

Beat together shortening and honey, add the eggs and vanilla. Mix throughly. Add your dry ingredients and mix until blended. Drop by spoonsfuls on a greased cookie sheet. Bake 350° for 12-15 minutes.

Pooh Honey Brownies
(white)

3 eggs, beaten
1 C wheat flour
1/2 C oatmeal
1/2 C wheat germ
1/4 t. salt
1/2 C chopped dates
1/2 C chopped nuts
1/2 C raisins
1 C. honey

Bake at 350 ° for 15-20 minutes in two 9x9" pans. Cut into bars.

Zucchini Brownies

2 C zucchini
1/2 C oil
2 C whole wheat flour
1/4 C carob powder or cocoa
1 1/2 t. soda
1/2 C honey
1/3 C fructose
2 T vanilla
1/2 C chopped nuts
1/2 C unsweatened carob chips

This makes an excellent sheet cake. It is fantastic plain or without frosting. In a blender, cut up zucchini, add oil, and blend well. Mix all together with the remaining ingredients in a large bowl until well blended. Pour onto a floured cookie sheet. Bake at 350 ° for 18-20 minutes.

Debbie Wade

Honey Date Fruit Bars

(dried foods)

2 1/2 C wheat flour
2 t. baking powder
1/2 t. salt
2 C honey
1/2 C dried egg mix + 1 C water
1 t. vanilla
3 C dried dates reconstituted
2 C chopped nuts
powdered sugar

Sift dry ingredients together. Set aside. Mix honey, eggs, and vanilla. Beat well. Add dry ingredients, reconstituted dates, and nuts. Spread in a 9"x13" pan. Bake at 350 ° for 45 minutes. Cool and sprinkle with powdered sugar.

Lisa Conn

Tuti Fruiti Bars

1 C. flour
1/2 C. butter
1/2 t. salt
3 eggs, beaten
1 C. chopped dates
2 T. sugar
1/2 C. nuts
3 T. flour
1 1/2 C. brown sugar
1 tsp vanilla

Mix flour, sugar and butter, pat into a 9 x 13 pan. Mix together remaining ingredients and pour over crust. Bake at 350° for 30 minutes. When cool , ice with a thin layer of butter cream icing.

Whole Wheat Chocolate Chip Bars

1/2 C margarine
1 C oil
2 C brown sugar
4 eggs
1 t. vanilla
1 t. salt
2 C whole wheat flour
2 C quick oats
1 C nuts
1 C chocolate Chips

Mix in order listed. Cream margarine, oil, sugar, and eggs together first. Mix all ingredients well. Bake in a greased cake pan at 350 ° for 30 minutes.

Connie Gardner

Congo Bars

2 C. brown sugar
2/3 C. oil
1/2 t. salt
1 sm. pkg chocolate chips
1 t. vanilla
3 eggs
2 3/4 C. flour
3 t. baking powdered
1 C. chopped nuts

Beat well sugar, eggs, and oil. Add remaining ingredients and mix well. Pour into a well greased 9 x 13 pan. Bake at 350° for 25 minutes.

Prize Winning Brownies

1 C sugar
1 C whole wheat flour
2 squares melted chocolate
1/2 C canned milk
2 T. butter
1 egg
1/4 t. salt
1 t. baking powder
1 C nuts
1 t. vanilla

Cream together sugar, butter, and vanilla. Add eggs and beat well until light and fluffy. Add chocolate, dry ingredients and canned milk. Mix well and then stir in the nuts. Pour into a greased 9x9 " pan. Bake at 350 ° for 30-35 minutes.

Mary Lois Madson

Honey Wheat Brownies

(pioneer recipe)

1/2 C butter
2 t. vanilla
2/3 C honey
2 lg. eggs
1/2 C wheat flour
1/2 C cocoa
1/2 t. salt
1 C chopped nuts

Cream butter, honey, and vanilla. Add eggs and blend well. Combine flour, cocoa, and salt. Gradually beat into creamed mixture. Stir in nuts. Put in a greased pan and bake at 350° for 25-30 minutes.

Pioneer Pastry

(150 year old recipe)

2 C shortening
1 t. salt
5 C flour
2 T vinegar
2 eggs
1 C cold water

Cut shortening into flour and salt until it is the size of peas. Beat eggs and vinegar together. Add this mixture to the flour mixture and pour cold water a little at a time into the center. Stir. Add only enough to hold the dough together and form a ball. Roll out dough. This recipe makes 2 double crust pie shells. Too much water or too much mixing will toughen the dough.

Hint For a lovely top pie crust finish , mix 1 egg yolk and equal amount of water. Beat until foamy, then spread a thin layer of foam over the top crust. Bake. The white of the egg works well to seal the edges of the pie crust together.

Peggy Layton

Whole Wheat Graham Cracker Pie Crust

1/2 C margarine
1 T brown sugar
1 C wheat flour
1/2 t. salt

Mix all ingredients. Press mixture into the bottom of pie pan. Bake at 350 ° F. for 15 minutes. This makes 1 pie shell.

Cherie Harmon

Lemon Pie Filling

1/4 C cornstarch
1 C sugar
1/3 C lemon juice
2 T. butter
1/8 t. salt
1 1/2 C boiling water
2 egg yolks

Mix cornstarch, salt, and sugar in a saucepan . Add boiling water and lemon juice. Cook until thickened, stirring constantly. Slowly stir the egg yolks and cook for 1 minute more. Add butter and let cool. Pour into a baked shell. You may top it with meringue.

Meringue: Beat 2 egg whites into stiff peaks. Add a dash of salt, and 1 t. vanilla. Put on the pie, and seal it over to the edge. Bake at 450 ° for 10 minutes or until lightly brown.

Judy Anderson

Chocolate Fudge Pie

1 C cocoa powder
2 C sugar
1/2 C dried egg + 2/3 C water
dash of salt
1 C margarine
1/2 C flour
2 t. vanilla
pastry

Melt margarine. Remove from heat and add cocoa powder and sugar. Beat the egg mixture with water until light. Add to the cocoa mixture and beat. Sift in flour, salt, and vanilla.

Pour into unbaked pie shell and bake 20 minutes at 350 °. Serve with whipped topping.

Jill Hanson

Golden Brown Fruit Pies
(Dried Foods)

pastry shell
4 C water
2 C dried fruit
1 C sugar

Simmer fruit in water until tender. Add sugar and thicken like pie filling. Roll out pastry and cut into 4 " squares. Place 2 T. filling in the center and fold two corners to the center and moisten with water. Fry in hot oil until lightly brown.

Lisa Conn

Rhubarb Cream Pie

1 C honey
5 T flour
1 T. margarine
4 C diced rhubarb
1/2 C sugar
1/2 t. nutmeg
2 beaten eggs
Pie crust

Combine honey, sugar, flour, nutmeg, and margarine. Add eggs and beat smooth. Pour over rhubarb in a 9" crust. Bake 10 minutes at 450 ° F., then another 30 minutes at 350 °.

Cherie Harmon

Sunday Dinner Apple Pie

(Dried Foods)

pastry shell
1 2/3 C dried apple slices
2/3 C sugar
1/4 t. salt
1/2 t. nutmeg
2 1/2 C water
2 T. cornstarch
1 t. cinnamon
3 t. lemon juice

Prepare pastry. Combine all ingredients in a saucepan and bring to a boil. Boil for 1 minute. Pour into a pastry lined pie pan. Cover with remaining pastry. Bake 425 ° F. for 35-40 minutes until gently browned.

Lisa Conn

French Apple Pie

Pie filling

3 C dried apple slices
5 C water
3/4 C sugar

Crumb topping

1 C flour
1/2 C margarine
1/2 C sugar

Pour apple mixture into an unbaked pie shell. Sprinkle with cinnamon. Top with crumb topping and bake at 375 ° for 45 minutes.

Elaine Westmoreland

Cherry Pie Filling
(Dried Foods)

3 C dried cherries
1/4 C cornstarch
1/2 t. salt
2 C water
1 C sugar
red food coloring

Add cherries to the water. Stir in cornstarch. Heat to boiling. Add sugar and salt. Stir until sugar is dissolved. Use in pie, cobbler, and desserts.

Connie Gardner

Peach Pie
(Dried Foods)

2 C dried peach slices
2 C sugar
4 t. lemon juice
Pastry shell
4 C water
1/2 C corn starch
1 t. salt

Combine water, peaches, and cornstarch. Bring to a boil, then add remaining ingredients. Stir until sugar is dissolved. Pour into a pie shell, cover with a crust. Bake at 425 ° for 35-40 minutes.

Jill Hanson

Thanksgiving Pumpkin Pie

1 3/4 C cooked, mashed pumpkin
1 3/4 C milk
1/4 C dried egg mix
1/3 C water
1 1/2 t. cinnamon
2/3 C brown sugar
1/2 t. ginger
1/2 t. nutmeg
1/4 t. cloves
1/2 t. salt

Mix all ingredients well. Pour into an ungreased pastry shell. 425 ° for 45-55 minutes. Serve with whipped topping.

Lisa Conn

Puddings were the most common dessert in pioneer times, especially bread pudding. As the fruit trees began to produce, pies, pastries, and other desserts became more common. Even the earliest Saints enjoyed pies on special occasions. Their pie fillings were made from wild berries found in the mountains and along streams. They would gather them all fall and dry them to make sure they had enough for their Christmas dinners.

Bachelor's Pudding

(Pioneer Recipe)

2 C cooked apples
2 C soft bread crumbs
1 egg, beaten
dash nutmeg
1 C raisins
1/2 C sugar
1 1/2 T. butter
1 t. lemon rind

Combine all ingredients. Spoon into a mold and steam 2 hours. Serve with cream or favorite sauce.

Elta Alder

Indian Pudding

(Pioneer Recipe)

5 C milk
1/2 C molasses
1 t. ginger
1/3 C cornmeal
1 t. salt

Cook the milk and meal in a double boiler for 20 minutes. Add molasses, salt, and ginger. Pour into a buttered pan and bake 2 hours in a slow oven, about (250 ° F.). Serve with cream.

Mary Fehlberg

Bread and Butter Pudding

Pudding

4 slices of buttered bread
2 eggs, beaten
1/3 C sugar
1/4 t. salt
3 C milk

Sauce

1/3 C butter
1 C pd. sugar
1/2 t. vanilla

Put the buttered bread in the bottom of a buttered baking dish. Mix remaining ingredients and pour over bread. Let stand 30 minutes. Bake 1 hour at 325 °. Cover the first half hour of baking.

Variation: Sprinkle 3/4 cups raisins or 1/2 C coconut between layers. Serve with the sauce.

Whip the sauce until fluffy and smooth.

Ruth Scow

Sweet Soup Pudding

(Pioneer Recipe)

This recipe came from the pioneers who crossed the plains with the Handcart Company.

Start with one quart of fruit juice. Be sure to drain all the fruit or put it through a sieve. Any fruit will work. The favorite is raspberries, cherries, dewberries, and black caps.

Boil the juice, mix 1/4 C cornstarch with 3/4 C water. Add a little of the hot juice to the cornstarch. Mix until smooth. Add the juice and cook until thick and smooth. Stir constantly.

Elta Alder

Indian Pudding

(Pioneer Recipe)

Pudding

1 egg
1/2 C bread crumbs, soft
1 1/4 C milk
1/4 t. cinnamon and nutmeg
1 T. honey
1 T. sugar
dash of salt

Sauce

1 C sugar
3 C water
1 T. butter
salt, vanilla, cinnamon

Beat egg. Combine with other ingredients. Put in a pudding dish or custard cups. Set in a larger pan of hot water. Bake slowly at 300 ° F. for 25 minutes. Serve with sauce.

To make the sauce, melt the sugar in a pan. Add 3 C water and dissolve by heating to boiling. Thicken it with a flour paste made from 1/4 cup flour and 1/4 cup water. Add salt , vanilla, cinnamon, and a T. butter to taste. Pour over pudding.

Peggy Layton

Corn Meal Indian Pudding

(Indian Recipe)

8 C milk
4 eggs
3 C raisins
nutmeg, cloves, and cinnamon to taste
4 T. cornmeal
2 1/2 c sugar
1 t. salt

Cook cornmeal slowly in the milk until it appears creamy. Add sugar and raisins to the beaten eggs . Add spices last. Open little places in the pudding and pour in the cream. Cook slow for 4- 6 hours until done.

Mery Fehlberg

Mormon Pioneer Rice Pudding

2 C rice
4 C water
3 eggs
4 C cream or fresh milk
1 C sugar
1/4 t. nutmeg
1/4 t. cinnamon
1/2 t. lemon extract
1 C raisins
butter

Cook rice with water until tender. Drain and add eggs, beaten cream, sugar , flavoring, spices, and raisins. Dot with butter in a glass baking dish. Cook at 300 ° F. for 45 minutes or until done when a knife inserted into the center comes out clean, and top is golden brown.

Peggy Layton

Danish Rice Pudding

3 C cooked rice
4 C warm milk
2 t. cinnamon
1/2 C raisins
1 C sugar or honey
2/3 C dried egg mix
1/4 t. salt

Combine rice, milk, sugar, and salt. Cook over low heat until thickened. Stir often. Add vanilla and raisins. Stir well. Rehydrate egg mix in 1/2 cup warm water. Beat until foamy. Remove pudding from heat and fold in eggs. Sprinkle with cinnamon and serve.

Elaine Westmoreland

Basic Pudding Mix

1 1/2 T. cornstarch
1 C milk
2 T. sugar
1/2 t. vanilla

Blend starch with sugar and 1/4 C cold milk. Scald remaining milk in the top of a double boiler. Gradually add the cold paste. Stir constantly. Cook over low, direct heat for 2 minutes. When pudding boils and thickens, cover with a lid and place over simmering water in the double boiler for 5 minutes. Add vanilla . Cool. If chocolate pudding is desired, add 1 T. sugar, 1 1/2 t. cocoa, and 1 t. margarine to the scalded milk.

Cathie Call

Best-Ever Apple Pudding

1 C. sugar
1/4 C. soft butter or margarine
2 large, unpeeled, shredded apples (2 C.)
1 C. sifted flour
1 tsp. baking soda
1 tsp. cinnamon
1/2 tsp. nutmeg
1/4 tsp. salt
1/2 C. chopped nuts

Cream sugar and butter. Add egg; beat well. Shred apples medium fine and add at once to creamed mixture. Stir in sifted dry ingredients. Add nuts. Bake in greased 8 or 9 inch square pan for 45 minutes at 350° . Serve with ice cream, favorite sauce, or whipped cream.

From the Mormon Colonies

Popcorn Pudding

(Fannie Farmer calls it simply "Corn Pudding", and suggests that one serve it with maple syrup or cream)

2 C. popped corn
3 C. milk
4 T. butter, melted
3 eggs, beaten
1/2 C. brown sugar
1 tsp. vanilla extract
1/2 tsp. salt

Grind all but a small handful of the popped corn in a food processor or grinder. Scald the milk, pour it over the corn, stir in the butter, and let the popcorn sit, covered, for 1 hour to absorb the liquid.

Beat eggs with the sugar until light, add vanilla and salt, beat in the corn mixture, and turn into a buttered baking dish. Sprinkle the reserved popcorn on top. Bake at 300° . until custard is set and browned on top, (45 minutes to 1 hour.) Serves 4 to 6.

Molasses was the main sweetener in the early years of Manti, as in all pioneer communities. Sugar cane was an important part of the farmer's crops. Before 1876 a tract of land to the south of Manti was employed for this reason and a molasses mill was operated to extract the sweetener. Candy was made from the molasses skimmings. The children enjoyed the taffy pulls that were held after each molasses harvest. The molasses was a coveted item. It was used for medicinal purposes as well as a sweetener.

Mormon Horehound Candy
(Pioneer Recipe)

2 oz dried horehound
3 1/2 pounds brown sugar
1 1/2 pints of water

Boil horehound and water for 1/2 hour. Strain it and add 3 1/2 pounds of brown sugar. Boil until hard ball stage. Pour onto a greased slab. When cool enough, cut into small pieces.

Peggy Layton

Honey Candy
(Pioneer Recipe)

2 C honey
1 C cream
1 C sugar

Combine and cook slowly until it reaches hard ball stage when tested in cold water. Pour onto a buttered platter. When cool enough to handle, butter hands and pull until a golden color appears. Pull into long ropes, and cut into pieces.

Vicki Tate

Old Fashioned Hardtack

2 C sugar
3/4 C karo syrup
1 C water

Cook to 270 ° or hard brittle stage with threads in the water. Remove from the heat and add food coloring as desired, and 1/2 t. flavoring, cinnamon, peppermint, etc. may be purchased in any drug store. Pour into a buttered dish and cool. Break into pieces and roll in powdered sugar.

Cherie Harmon

Honey Peppermints

1 C warm honey
4 drops of oil of peppermint
green or pink food coloring
2 3/4 C pd. milk

Mix ingredients and knead until all the milk is absorbed. Pull like taffy. Then stretch into rows and cut.

Elaine Harmston

Whole Wheat Candy

1 C butter
1 1/2 C whole wheat flour
nuts, coconut, sesame seeds
1 C honey
1 C peanut butter

Melt butter, honey, and peanut butter. Add flour. Cook and stir a few minutes. If you don't like raw wheat, add nuts, coconut, or sesame seeds.

Judy Anderson

Old Fashioned Honey Carmels

2 C honey
1 can evaporated milk
3 T. butter
1 C chopped nuts
dash of salt
1 t. vanilla

Mix honey and milk and cook until it forms a firm ball (255 °). Stir in butter, nuts, and salt. Pour onto buttered dish. Cool and cut.

Peanut Butter Oatmeal Log Roll

2 C rolled oats
2 C powdered milk
1 C raisins
1 C peanut butter
1 C corn syrup

Combine rolled oats and peanut butter. Mix well. Add remaining ingredients. Using hands, mix well, separate into 4 parts. Roll into the shape of a log. Slice into 1/2 " pieces.

Elaine Harmston

Granola Candy

Honey, coconut, and granola.

Add enough honey and coconut to the granola to make it stick together. Form into balls and wrap in wax paper. For a nice variation add chocolate chips.

Honey Peanut Butter Nuggets

1 C peanut butter or
1 C peanut butter powder reconstituted
1 C powdered sugar
1 C honey
1 C powdered milk

Mix all ingredients until well blended. Roll into 1 " balls. Refrigerate until firm.

Elaine Harmston

Summer Popsicles

1 pkg jello
1 C sugar
1 pkg kool aid
2 C hot water

Mix together until dissolved. Add 2 cups of cold water and freeze into popsicle molds or ice cube trays with sticks in them.

English Toffy

1 C sugar
2 squares margarine
1 T. water
grated chocolate
chopped nuts

Combine sugar, margarine and water in a heavy pan. Cook on high. Stir constantly until it turns a light tan color. Remove and pour into a pan. While still hot, top with grated milk chocolate and chopped nuts.

Peggy Layton

Peanut Brittle

2 C sugar
1/2 C water
1 C light corn syrup
4 T. butter
2 C raw peanuts
1/8 t. salt
1 t. baking soda
2 t. vanilla

Combine all ingredients in the first column in a heavy saucepan. Bring to a boil, cook to 242 ° F. Add peanuts and cook until mixture reaches 294 ° F. Stir constantly. Mixture will be quite thick. Remove from heat and add baking soda and vanilla. Mixture will foam. Pour onto a greased pan. When thoroughly cooled, crack into pieces.

Peggy Layton

Pudding Fudge

1 C sugar
1 pkg. chocolate pudding
1/2 C canned milk
2 T. butter

Mix all together except butter. Cook until soft ball stage. Mix in butter. Add nuts if desired. Pour into a buttered dish. Cool and cut into squares.

Cathi Call

Lollipops

2 C sugar
1 C water
3/4 C corn syrup
flavoring and coloring

Boil to hard ball stage. Add flavoring and coloring as desired. Pour out on sticks on waxed paper.

Lisa Conn

Homemade Marshmallows

2 env. Knox gelatin
1 C boiling water
a pinch of salt
4 T. cold water
2 C sugar
1/2 t. flavoring

Place gelatin in a bowl with 4 T water. Stir to soften. Add boiling water, sugar, salt, and flavoring. (vanilla can be used). Beat with a mixer until thickened. Approximately 10-15 minutes or just before it peaks. Fill cake pans with 2-3 inches of flour. Make imprints with an egg for molds. Fill with marshmallow mixture.

Vicki Tate

Trail mix

1 C raisins
1 C dried apricots
1 C chocolate chips
1 C nuts
1 C dried apples
1 C dried peaches
1 C sunflower seeds
2 C granola
***Fruit Galaxy can be used in place of dried fruits.**

Mix all together and store in plastic ziplock bags. This is a great snack. It is good for camping and backpacking.

Honey Caramel Popcorn

1 C. honey
2 C. sugar
1/4 C. margarine powder

Bring to a boil for 2 minutes. Pour over popped, salted popcorn.

Jello Popcorn

1 C. light corn syrup
1/2 C. sugar
1 (3-oz.) pkg Jello
9 C. popped corn

Bring syrup and sugar to a boil. Remove and add Jello. Stir until dissolved. Coat popcorn and form into balls.

Cracker Jacks

1/2 lb. popcorn - popped and kept hot

In a large pan, mix together and cook to a soft ball stage:
1 1/2 C. sugar
1 C. brown sugar
1/2 tsp. cream of tartar
2 T. dark corn syrup
3/4 C. water
Add: **1 C. raw unsalted peanuts** and cook to a crack stage.
Add: **1 tsp. salt** and cook to a hard crack stage
Add: **1/2 C. melted butter**
1 tsp. baking soda
1 tsp. vanilla

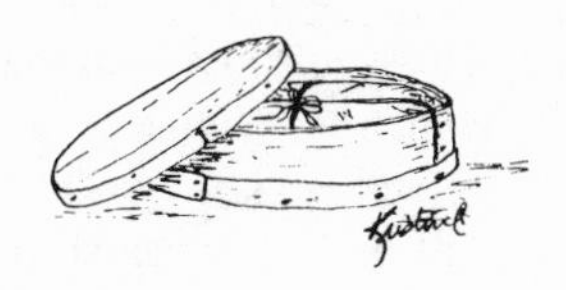

Pour over corn and stir well.

Sugar and Spice Popcorn

2 qts. popped popcorn
3 T. butter
2 T. sugar
1/2 tsp. cinnamon

Melt butter, add sugar and spice and stir until sugar is dissolved. Drizzle over popcorn and toss.

Molasses-On-Snow Candy

1 C. dark molasses
4 9-inch pie pans
6-oz. glass
1/2 C. brown sugar
2 qt. saucepan
1-cup heat-proof pitcher

Fill pans with fresh snow, then set outside in snow to chill when you start cooking. Combine molasses and brown sugar in the saucepan (don't use a smaller one; candy needs to boil up) and bring to a boil. Continue to cook on medium heat , stirring frequently to prevent burning.

After 5 minutes begin testing the syrup by dripping some from a spoon into a glass of cold water. The drops may dissolve and make the water cloudy, or they may form soft balls. Using clean water each time, repeat test every few minutes until a sample forms a firm ball in the water (250°). Remove from heat. Pour half the hot candy into the pitcher so two people can pour two portions.

Fetch pans of snow. Working rapidly, pour hot syrup onto cold surfaces. When candy has hardened break it into bite-size pieces, setting two aside for the cooks.

Molasses was the main sweetener in the early years of Manti, as in all pioneer communities. It was made from sugar cane, which was an important part of the farmer's crop. Before 1876, a tract of land to the south of Manti was employed to grow this crop and a molasses mill was operated to extract the sweetener which was a coveted item in that day. Candy was made from the molasses skimmings and the children particularly enjoyed the taffy pulls that were held after each molasses harvest. As well as being used as a sweetener, molasses was also used for medicinal purposes.

Almond Joy

1 C powdered milk
1 C coconut
1/4 t. salt
1 recipe of honey carmels
1/2 C honey
1 t. vanilla

Mix honey, milk, salt, and vanilla together. Add coconut and press mixture in a small square pan. Add 1 C chopped, toasted almonds on top. Pour a layer of hot honey carmel over all the candy. Cool and cut into small squares or bars and dip into chocolate.

Lisa Conn

Peanut Butter Kisses

1/3 C karo
1/3 C dry milk
1/3 C peanut butter
1/3 C powdered sugar

Mix syrup and peanut butter together. Gradually stir in milk and sugar. Roll into a roll 3/4 " in diameter, or you can roll it in chopped nuts. Chill then cut into 1 " pieces.

Vicki Tate

Carob Fudge

2 C peanut butter
1 C seasame seeds
1/2 C carob powder
1/2 C honey
4 t. vanilla

Mix peanut butter, honey, carob, and vanilla together. Add seeds. Knead together. Form into balls and roll in finely ground unsweetened coconut or chopped nuts. Excellent healthy, energy snack.

Debbie Wade

Toffee Squares

1 C shortening
3/4 C brown sugar
2 eggs
2 t. vanilla
4 C flour
1 t. salt

Cream all ingredients. Spread on a cookie sheet that is greased. Bake at 350 ° for 15-20 minutes. Remove from oven and sprinkle 1 C chocolate chips and spread until melted. Press in 1/2 C. nuts or coconut. Cool and cut into squares.

Anna Jean Hedelius

Carmels

2 C sugar
1 1/2 C corn syrup
1 t. vanilla
1 1/2 C cream
1/2 C butter

Combine all ingredients together and cook to the soft-ball stage. Stir occasionally. Pour mixture onto a well buttered dish or pan. Cut into squares and wrap in plastic.

Peggy Layton

Peanut Butter Cups

1/2 C peanut butter
1/8 C Karo Syrup
1/4 C butter or margarine
1/4 C. pd. milk
dipping chocolate

Knead together, all ingredients except dipping chocolate. Melt chocolate separate. Put a layer of chocolate in paper cups, then a layer of peanut butter, then more chocolate.

Vinegar Taffy

1/2 C vinegar
2 C sugar
1/2 C. water

Combine and boil to brittle stage. Pour onto a buttered pan. Stretch.

Honey Butter

1/2 lb. butter, softened

3 T. honey

Mix honey and butter together until well blended. Pack into a crock or serving jar and chill. Use on pancakes, waffles, hot breads, and french toast.

Vicki Tate

Fruit Butter
(Dehyd. Food)

2 lbs. dried fruit
2 C. sugar
6 C. water

Simmer dried fruit in water for 30 min. Stir in sugar. Cover and simmer 30 min. longer. Fruit should be soft. Place in blender and puree or put through sieve. Store in covered jars in refrigerator. Great as a filling for cookies, muffins, or spread on toast or pancakes.

Vicki Tate

Toppings for Breads or Pancakes

#1 Orange Honey Butter
- **2 T. honey**
- **1/2 C. soft butter**
- **2 T. Tang or orange juice concentrate**

#2 Honey Nut Butter
- **4 T. softened butter**
- **2/3 C. honey**
- **1/4 C. walnuts**
- **cinnamon if desired**

#3 Date-Honey Spread
- **2 T. honey**
- **2 T. butter**
- **1/2 C. dates or raisins, chopped**
- **2 small pkgs. cream cheese**

Cherie Harmon

Banana Whipped Topping

1 egg white, beaten stiff
1 t. sugar
1 banana, mashed

Beat egg white, add mashed banana, 1 teaspoon at a time, beating constantly. Add sugar and beat until light. Serve on fruit or pudding.

Vicki Tate

Chocolate Sauce

2 C. sugar
4 T. cocoa
1 can evaporated milk

In a saucepan over high heat, stir sugar and cocoa for 2 minutes, with a wooden spoon. Stir in evaporated milk and continue stirring constantly until you get a full rolling boil, turn off heat and continue stirring until boil subsides. Then serve over your favorite ice cream.

Honey Syrup

3 C. honey
3 C. water

Combine and bring to boil, stirring to blend. Simmer gently 5 min. Serve over pancakes and waffles.

Susan Westmoreland

Wild Chokecherry Syrup

1 Qt chokecherry juice (Boil berries 25-30 min. Put through collendar to get the juice.)
4 C. sugar
1 pkg. Knox gelatin

Boil juice and sugar together about 10 min till starts to thicken. Add gelatin. Fill bottles. Use wax to seal. (If you don't have gelatin, omit.)

Elta Alder

Strawberry or Raspberry Sauce

1 C fresh strawberries or raspberries
1/2 C water
1 1/2 T.
1 T. lemon juice
sugar to taste

Add water to fruit. Bring to a boil. Put the fruit through a sieve and extract the juice . Add enough water to make 1 1/2 cups juice. Add the cornstarch to a small amount of water, then add it to the juice. Cook until thick and clear. Add 1 T. lemon juice and sugar to taste.

Peggy Layton

Maple Syrup

Bring to boil and cook 1 minute
1 3/4 C. white Sugar
1/4 C. brown sugar
1 C. water
Add:
1/2 t. vanilla
1/2 t maple flavoring

To help syrup not crystalize as it stores, cover saucepan as it cools down.

Jill Hanson

Fruit Syrup for Pancakes
(Fresh or dried)

Bring to boil, stirring constantly:
1/2 C. sugar
3 T. cornstarch
2 C. water (use fruit juice for part of water)
Add:
2 C. sliced peaches or other fruit. Simmer till fruit is tender. Add 2 T. lemon juice. Serve hot with pancakes.

Vicki Tate

Pineapple Sauce for Pancakes

Melt in sauce pan:
3 T. margarine
Add:
1 C. crushed pineapple
2 T. brown sugar

Heat 5 min. stirring until clear and thick. When sauce cooks down a bit, add a dash of nutmeg. Serve hot or cold on pancakes, waffles, or ice cream.

Vicki Tate

Wild Berries for Pies and Jam

Pioneers used:

Ground Cherries (fruit of a weed)
Choke Cherries
Service Berries
Gooseberries

All were gathered wild and used for jam or pies. The ground cherries are sour like pie cherries and when gathered at end of summer they ripen by Thanksgiving for pies.

Elta Alder

Herseys Chocolate Syrup

4 T. baking cocoa
3 T powdered sugar
1/4 C water
small bottle of corn syrup

Make a paste of the cocoa, water, milk, and sugar. Add to the bottle of corn syrup and mix well.

Elaine Harmston

Butterscotch Sauce

1 1/2 C brown sugar
2/3 C heavy cream
2/3 C corn syrup
1/4 C butter

Mix all ingredients in a saucepan. Bring to a boil over medium heat. Boil 3 minutes and cool.

Peggy Layton

Basic Sweet Sauce

Prepare ahead

1/2 C cornstarch	**4 C sugar**
1 C margarine	**1/2 t. salt**

Blend until mixture is uniform and crumbly. Store this in a jar in the refrigerator. Label it. Pack the mix as you would brown sugar. Blend 1/3 C. mix with 2/3 C. cold liquid. Season as indicated below. Heat to boiling, stirring constantly.

Cherry sauce: Add juice from red sour cherries and water to make 2/3 C. Add 1/2 C red cherries and red food coloring after mixture has thickened. (Optional) : add 1/4 t. almond extract.

Orange sauce: Use orange juice to replace the cold liquid.

Lemon sauce: 3 T. lemon juice and water to make 2/3 cup liquid.

Raisin sauce: Water and 1/4 C. raisins.

Chocolate sauce: Milk and 1/2 square of chocolate.

Pineapple sauce: Juice from crushed pineapple and water to make 2/3 C. Add 1/2 C crushed pineapple.

Strawberry sauce : Add 1/2 C fresh strawberries and a few drops of `red food coloring.

Cathi Call

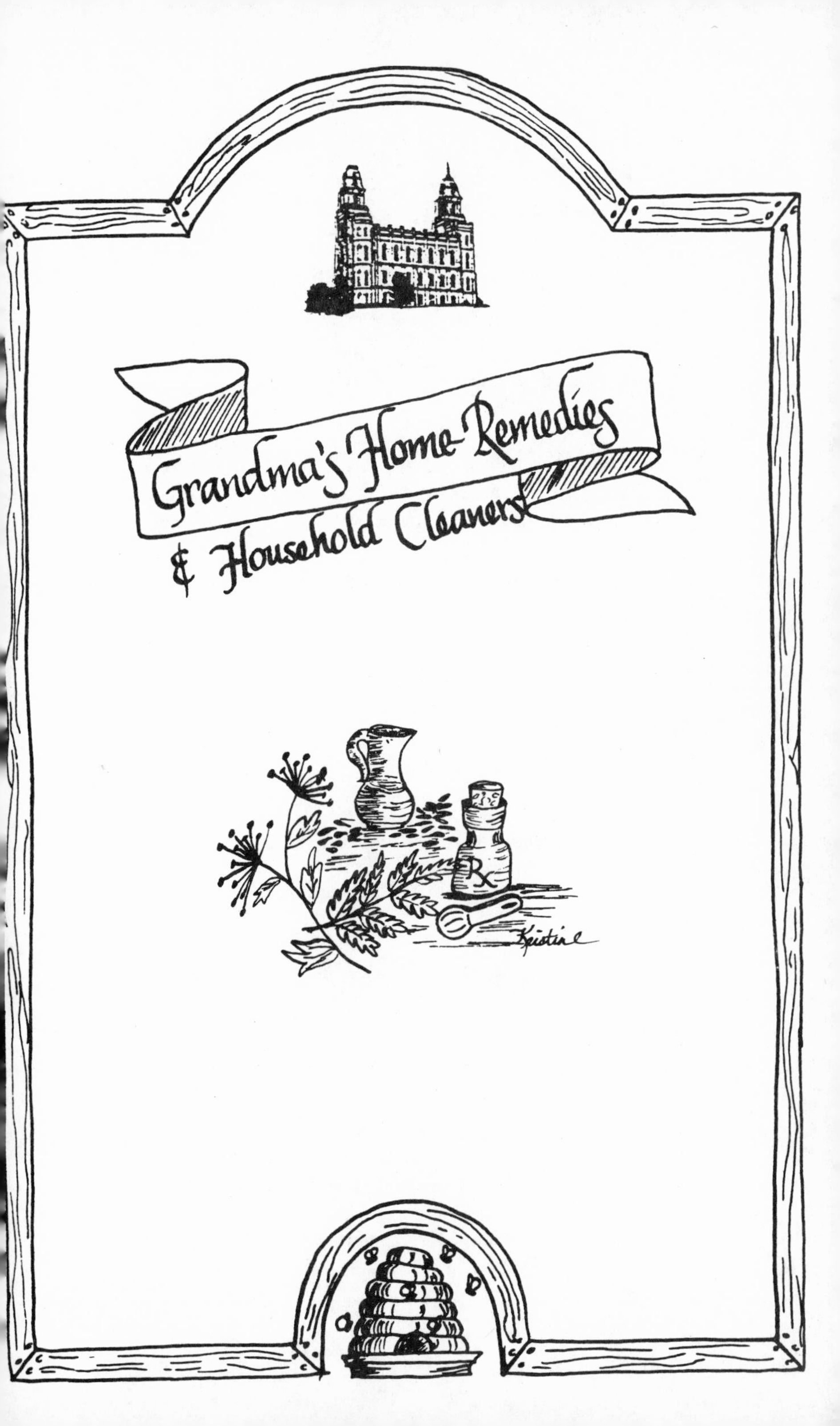
Grandma's Home Remedies
& Household Cleaners

Children's Soap Bubbles

2/3 C. water
1/3 C. Joy or Dawn liquid dish soap
3/4 tsp. glycerin
food coloring

In a bowl, stir together the water, detergent, and glycerin. Add the food coloring drop by drop to obtain desired color. Pour mixture into an empty bottle of commercial bubbles and use the accompanying plastic hoop.

Homemade Playdough

1 C flour
2 t. cream of tarter
1 C water
1/2 C salt
1 T. oil
food coloring

Cook all ingredients slowly, stirring until it forms a ball. Cool, and knead until lumps are gone.

Andrew and Angela Tate

Cornstarch Fingerpaints

2 C water
1/4 C cornstarch
food coloring

Bring water to a boil. Add a small amount of water to the cornstarch and blend together to make a paste. Gradually add paste to the boiled water. Stir well. Bring to a boil, remove from heat, and add food coloring. Cool and play. Use butcher paper. You can also paint with pudding. It's fun and you can lick your fingers.

The wild berries and herbs growing in the mountains and valleys of Sanpete county were a wonderful blessing to the early saints. It took several weeks to make the treacherous journey by covered wagon to Salt Lake City for medical care. They often were too sick to travel and just died. If the saints were sick, they had to rely on the herbs that they could find growing wild. Molasses was used as a medicine, along with the soothing teas made from the wild herbs.

Doc Smith's Cough Syrup

1 C honey
juice of 2 lemons
1/2 C hot water
1 T glycerin

Mix well together and take 1 tablespoon at a time when a cough persists.

Cough Remedy

2-3 garlic buds pressed and warmed in 2-4 T. honey. Add 1 t. cayenne pepper. Take 1 t. at a time for cough.

Marilyn Harris

Ear Ache Treatment

A heated onion, warm to the touch cut in half is a great poultice, especially for an ear ache.

Marilyn Harris

Mosquito Bites

1/4 C alcohol
2 aspirins

Shake until it dissolved. Apply to the bite and let dry.

Tooth Paste

Soda on a toothbrush works well to clean teeth. Rinse the brush with hydrogen peroxide. This prevents plaque and gum disease.

Vitamin C

Just after the frost, gather a lot of clean rose hips. Snip off tails, and spread on a clean surface. Let them dry partially. When the skins begin to feel dry and shriveled, split the hips and take out the large seeds. If they are not dry enough the seeds will be sticky and cling to the pulp. If they are too dry it will be hard to remove the seeds. After the seeds are removed, allow the hips to dry completely. Very nutritious and extremely high in Vitamine C. These make a good snack.

Ointment for Frost Bite

1 C vinegar
1/2 C salt
ice water

Apply to affected area. For sprains or strained muscles, use hot as a compress.

Ruth Scow

Slivers

Use a bottle of household turpentine. Dab the sliver with it . It will make the skin turn clear so you can see the sliver. Then squeese the area around the sliver and grab it with tweezers and pull it straight out. If it won't come, dab more turpentine on it and it should come.

Sore Throat

To make a gargle, use 1/2 cup of warm water and 1/2 teaspoon of salt.

Cracks in Feet or Toes

Wool is a healer of feet. A wool rug will help. Take some white sheeps wool or wool yarn, put a tiny bit of vaseline on the wool and tie it around the toe. It will heal fast.

Poultice for Cuts and Bleeding

Comfrey is the number one healing herb. In any herb poltice, use four times as many fresh leaves as dried ones. Chop the leaves finely, soak them in boiling water and put it on the wound. Let them cool sonewhat first so you don't get burned. When this one cools down, put a different warm one on. Do not reuse the herbs. Keep a clean cloth around the poltice so it will stay on.

Flea Protection

If you love animals, there is always a chance of getting fleas in the carpet and bedding. Sprinkle talcum powder into the animals fur and rub it down to the skin. The powder smothers the fleas. Be sure to cover the animal with it, until a cloud of powder comes up when the fur is patted. A bay leaf under the sheets of a mattress will repel fleas.

Bee or Wasp Sting

Apply a small amount of Adolph's Meat tenderizer mixed with a little water to make a paste and apply to the sting. The papaya enzyme in the tenderizer neutralizes the bee poison.

Sunburn

Aloa Vera is known also as the burn plant. Take a tip off one of the leaves and squeeze out the salve that is inside the plant. Apply this to the burn. It is good for any kind of a cut or wound. It will heal it fast. The plant will repair its own cut also.

Itching Hives

A hot bath with baking soda will work wonders. Use1 cup of soda in a half full bathtub. Use 2 cups in a full bathtub.

High Fever

Give Tylonol to bring the fever down then put the child or adult in a cool tub of water. Continue this every 4 hours. Another remedy - tie strips of flannel dipped in vinegar around the patient's wrists. As strange as this sounds it's a tried and true fever reducer.

Nosebleed

Blow your nose hard first. This removes the clots, then sit up and pinch your nose for 8 minutes. Time it and don't be tempted to take the pressure off. Laying down only causes the blood to go down your throat faster.

Cuts, Scratches, and Infection

Wash the cut with soap and water, rinse with hydrogen peroxide. A dressing for the infection is Epsom salts in 1 cup boiled water. Soak a cloth with the solution and put it on the infection. Wrap it with a clean cloth.

Canker Sores

Sprinkle dry thyme on the canker sore. It should heal in a few hours.

Leg Cramps

More calcium in the diet is needed.

Croup in Babies

Chop the fresh garlic. First put a layer of vaseline on the bottom of the feet, then put a layer of thin cloth, and then a layer of garlic. Cover the poultice with a clean cloth. Bandage the whole thing to keep the poultice in place. Leave it all night. If the child seems to be strangling, quickly take wash clothes and saturate with rubbing alcohol. Fold once, lay it on the throat, and upper chest. Put a dry towel over it and hold in place. The child will breath freely in a few seconds. Put a humidifier in the room with the child. You can even make a tent around their crib or bed and put the steam into the tent so the child can breath it. If it continues call the doctor.

Dandruff

After cleaning the hair and rinsing, apply cider vinegar straight. Do not rinse it out. Rub it in well. The odor will disappear in a few hours. Repeat if needed.

Eye Infections

For eye infections, wash with sagebrush tea or golden seal tea.

Headaches

Catnip, peppermint, or garden sage tea. Take a brisk walk to relieve stress.

Hint Whenever a tea is mentioned as a home remedy the recipe is 1 oz. herbs to 1 pint of boiling water. Let it steep for about 20 minutes, then remove the herbs and drink.

Laxative

Eat whole wheat cereal or bread. Peach leaf tea, sage tea, and raw apples.

Indigestion or Heart Burn

Eat a raw potatoe. Drink sage or peppermint tea, or eat raw apples.

Hemorrhage

Drink a green drink made from alfalfa you can get at a health food store. Take 1 capsule comfrey, and 1 capsule cayenne every 3 hours.

Old Fashioned Mustard Plaster (for chest colds)

4 T. flour
2 t. oil
1 t. dry mustard

Mix in lukewarm water to form a paste. Spread on thin clean cloth and cover. Place on chest for 20 minutes.(Shorter time for small children) Be careful not to burn the skin. Remove the plaster and cover the chest with camphorated oil or Vicks. Then cover with warm fabric such as flannel or a towel. Repeat in 4 hours.

Grandma's Hand Soap

1 can lye
1/2 C ammonia
1/2 C powdered borax
4 t. aromatic oil of rose or lavender
3 oz. glycerine
3 T. finely ground oatmeal
11 C melted and strained fat
5 C rain or soft water
1/3 C sugar

Pour water into an enamel pan, never use aluminum or tin. Stir well after each of the following: Lye, ammonia, borax, and sugar. Stir until cool. Slowly pour in fat. Continue stirring as you pour. Add the fragrance and stir for 15 more minutes. While stirring add lanolin, glycerine, and oatmeal. The mixture should be thick and creamy. Pour into molds the size of bars you want. Let stand until firm. Wrap in wax paper. Let it age for at least 1 week before using.

Lisa Conn

Laundry Soap

5 lbs grease
1/2 C powdered borax
1/2 C coal oil or kerosene
1/2 C ammonia
1 can lye

Melt the lye in a quart of water. Dissolve borax in 1 cup of water and add to lye mixture. Melt the grease, add the ammonia and coal oil. Add the lye. Stir until it congeals. Pour into milk cartons. This soap can be ground or grated to make granulated, laundry soap. Do not use tin or aluminum container when making soap.

Jill Hanson

Note Make the soaps outside and keep all ingredients away from children.

Best Ever Window Cleaner

1 pint rubbing alcohol
1 T. liquid detergent
4 T. ammonia

Add enough water to make a gallon. Add a little blue coloring.

Household Cleaner

1/2 gallon water
1/2 C vinegar
1 C ammonia
1/4 C soda

Add enough water to make 1 gallon. Shake well. Use in a spray bottle.

Wallpaper Cleaner

2 C flour
2 1/2 T. ammonia
1 t. baking soda
1 1/4 C water

Stir soda into flour. Then add ammonia and water. Mix well and steam for 1 1/2 hours. Cover until cool, then knead in hands until it is like art gum. Rub the wallpaper, turning cleaner often to fold in the dirt.

Removing Oil From Wallpaper

Make a paste from fuller's earth and carbon tetrachloride. (Can be purchased at any drug store). Spread the paste on so as to cover the area completely. When dry, wipe off with a clean cloth. Use in a well ventilated room.

Vicki Tate

Rug Shampoo

6 T. Dreft
1 quart warm water
4 T. ammonia

Beat all ingredients until sudsy. Use with a brush or sponge to clean the rug or furniture. Wipe with a clean rag.

Ruth Scow

Furniture Polish

1 quart hot water
1 T. turpentine
3 T. boiled linseed oil

Keep in a covered can. Reheat to use. Rub on, wipe off, and polish.

Elaine Harmston

Drain Cleaner

1 C baking soda
1/4 C cream of tarter
1 C table salt

Put 1/4 cup of solution in the drain. Add 1 cup water. Let stand for a little while .

Silver Polish

Make a paste using wood ashes and water. It will really shine silver and copper. Wood ashes andwater makes lye.

Jewelry Cleaner

Soak tarnished jewlery in a solution of concentrated lemon juice.

Remove Labels from Jars

Use hot vegetable oil on the label. Let it sit for a while until the label peels off.

Emergency Candles

2 cans of fine sawdust **1 lb. melted wax**

Chip or cut up wax into small pieces, and melt over water, never directly over flame or burner. It can be melted in the same can you're going to use for the candle. Put the sawdust in a shoebox and pour the wax over it. Mix it until it holds together well when squeezed in your hands. Pack and press the mixture into the can tightly and firm. Make a hole with a long knitting needle. Add the wick, all the way down. Pour melted wax over the top to secure.

Flea Powder

If you love animals, there is always a chance of getting fleas in the carpet and bedding. Sprinkle **talcum powder** into the animal's fur and rub it down to the skin. The powder smothers the fleas. Be sure to cover the animal with it, until a cloud of powder comes up when the fur is patted. A **bay leaf** under the sheets on a mattress will repel fleas.

Charcoal

Charcoal is a very useful fuel. It can be made from **twigs and limbs of fruit, nut and other hardwood trees;** from **black walnuts or peach and apricot pits.** It makes a hot fire which gives off little or no smoke.

To make charcoal, simply put the wood in a can which has a few holes punched in it. Put a lid on the can and "cook" it over a hot fire. The holes in the can will allow the gases and flame to escape. The exclusion of oxygen keeps the wood from completely burning to ashes. When the flame from the holes in the can turns to yellow-red, remove the can from the fire and allow to cool. Store in paper bags or cardboard cartons.

Survival Foods

Flowers which can be Eaten

Apple blossom	Chrysanthemum	Primrose
Red clover	Daisy	Marigold
Dandelion	Marjoram	Rose
Broccoli	Daylily	Nasturtium
Calendula	Elderberryflower	Sage
Carnation	Forget-me-nots	Pansy
Thyme	Cauliflower	Geranium
Peony	Viola	Chamomile
Gladiolus	Pinks	Violet
Chives	Lavender	Plum blossom
Yucca	Rosemary	Lemon blossom

Poisonous Flowers

Do not Eat under any Circumstance

The following flowers should **not** be eaten. They are known to be toxic. Plants which have been known to cause death are marked with an asterisk(*). The plants on this list are by no means all of the poisonous plants, they are simply some of the most common ones. Always be sure of your botanical identifications. If you aren't quite sure of a plant's identification, admire it, but don't eat it!

Amaryllis	Anemone	Angel's trumpet*
Arnica	Autumn crocus	Meadow saffron*
Azalea*	Belladonna*	Bird-of-paradise
Bittersweet	Bleeding heart	Cape jasmine
Castor bean*	Cestrum	Cherry blossom*
Chinaberry*	Christmas rose	Clematis
Crownflower	Daphne*	Four-o-clock
Foxglove*	Glory lily	Golden chain*
Holly	Horse chestnut	Hyacinth
Hydrangea	Jessamine	Lantana*
Larkspur	Laurel*	Lily-of-the-Valley*
Monkshood*	Narcissus	Oleander*
Prickly poppy	Rhododendron*	Snow-on-the-mountain
Rhubarb Blossom*	Sweet pea	Tomato blossom
Trumpet flower	Wisteria	

Mary Fehlberg

Elta Alder, of Manti, tells an interesting story of her youth on the Green River in the early 1900's. She says that a band of Indians camped every year in her father's pasture during hunting season. The women tended camp and preserved the meat while the men spent days hunting deer. Every evening one of the braves would bring "the kill" to the camp where the women cut the meat into strips the size of a person's finger and soaked them in salt water over night. Each strip was then hung on a barb on the barb wire fence. By the end of the hunt, every barb around the entire field had a strip hanging from it. These were collected in sacks and stored to provide meat for the winter.

Emergency Baby Formula

1/3 C. plus 2 t. instant powdered milk
or
1/4 C. non instant powdered milk
1 1/3 C. boiled water
Mix together and stir thoroughly.
Add:
1 T. oil
2 t. sugar

If baby bottles are not available, milk can be spoon fed to an infant.

Emergency Baby Food

3/4 C. cereal grain
1/4 C. beans

Boil until soft, then press through a sieve. Boil again to insure that it is bacteria free.

This will provide good protein and iron, as well as calories. This can be fed to infants under 6 months if adequate milk is not available, but it must be pureed to a fine texture.

Vickie Tate

Emergency Survival Bar

3 C. cereal (oatmeal, cornmeal, or wheat flakes)
2 1/2 C. powdered milk
1 C. sugar
3 T. honey
3 T. water
1/2 C. jello (optional)
1/4 t. salt

Place all dry ingredients except jello in a bowl. Bring water, honey, and jello to a boil.. Add to dry ingredients. Mix well. Add water a little at a time until mixture is just moist enough to mold. Place in a small square dish and dry in the oven under very low heat. Wrap and store. This will make 2 bars, each containing approx. 1000 calories or enough food for one day. These will store for a long time and are excellent for emergency packs, etc. Eat dry or cooked in about 3/4 C. water.

Vicki Tate

Buffalo Jerky
(Pioneer Recipe)

Slice the buffalo meat along the grain into thin strips 2-3 inches long and 1/2 " wide. Soak it overnight in a salt water solution. Hang them on a barbed wire fence on each barb or on bushes to dry in the sun. They can be baked in a 200 degree oven until dry or suspended over a fire to dry. This method can be used for beef or venison jerky also.

Venison Jerky

Marination Sauce:

2 parts seven-up (1 qt.)
1 part soy sauce (1 pt.)
1/4 part Worcestershire (1/2 C.)
2 T. garlic salt

Cut venison in strips about 1/4 " thick. Put into cold salt water with about 1 T. soda. Soak about 20 minutes. Rinse in cold water until all of the blood is washed out.

Marinate over night, about 12-18 hours. Take out of marination and dry it off on paper towels.

Now you can add flavoring to the meat as desired.

1. You can rub brown sugar on the meat and sprinkle with black pepper.
2. For jerky with a "bite", put Tabasco sauce on the meat.
3. Sprinkle with seasoning salt, garlic salt, or onion salt, or anything else you desire.

Place meat strips on racks from the smoker. Place inside the smoker. It will take about 24 hours to smoke, dry and cook the meat. The length of time it is left in the smoker depends on whether you like it completely dry or a little spongy(not quite so well dried). The thickness of the meat also helps determine the length of time smoking.

After this process is completed, cut it up into small pieces and put into small plastic bags--It will go a lot further that way. If the jerky is not completely dry, you will probably need to freeze it.

MoRell Snow

Wheat Jerky

(Gluten)

2 C. raw gluten, rolled thin and cut into jerky-size strips
2 C. broth

Broth:

1/4 to 1/2 C. soy sauce
1 T. worcestershire sauce
1 T. honey
1/2 tsp. onion powder or salt
1 tsp. sausage seasoning (recipe below)
1 1/2 C. water
1 tsp. liquid smoke
1 tsp. black pepper
1/2 tsp. garlic powder or salt
1/2 tsp. tobasco sauce (optional)

Mix broth ingredients and bring to boil. Drop gluten pieces into broth, one at a time, and simmer 10-20 minutes, stirring occasionally.

Place pieces on cookie sheet that has been coated with a non-stick spray. Place in 200°-250° . oven (leave door ajar slightly) until pieces appear dry on top. Turn pieces over and sprinkle black pepper on if desired, then leave in oven until texture is firm and chewy, about 30-60 minutes.

Note Can also be dried in a food dehydrator or draped over a wooden stick and placed in hot sun for about 2 hours.

Will keep in jar on shelf for several months; or store in plastic bags in refrigerator or freezer.

Turkey Jerkey

Use sliced turkey pastromi, cut up into strips and lay flat in a food dehydeator or in the oven at 200° for 1 hour or until dry - Eat and enjoy.

EDIBLE WILDLIFE

Most animals and reptiles and many insects are edible, however, avoid any small mammal which appears to be sick as it may have tularemia, a disease transmittable to humans. A spotted liver in the animal is also an indication of this disease. Some animals have scent glands, which must be removed before cooking. Do not allow the hair of these animals to come in contact with the flesh as it will give the meat a disagreeable taste.

Locust

Gather locust at night, remove the shells. Do not let them be exposed to the sunlight or they will spoil. Wash and then fry in a small amount of hot oil. Eat and enjoy.

Elta Alder

Ground Hog

Skin and clean thegroundhog. Boil until tender. Remove from the water and season with salt, pepper, and some red pepper. Bakein an oven about 350 or cook over an open fire.

Elta Alder

Frogs and Toads

Grab the frog or toad by the neck, Twist off heads, scald and peel off the skin while running cold water over the frog. slightly boil, remove from the water and bake in the oven, fry, or cook on a roasting stick over an open fire.

Elta Alder

Raccoon

Clean and skin th raccoon, Boil in scalding water seasoned with red pepper. When it is tender, remove from the water. Add the salt, and pepper, and bake in the oven until brown. The raccoon can also be fried

Elta Alder

Birds or pheasant

Clean birds. Any bird can be cooked like chicken. Either boil until tender, or run a stick through it and roast before the fire. This is good served with rice or noodles or made into soup.

Elta Alder

Other Edible Wildlife

1. Jack Rabbit: Has long ears and legs, sandy color, and may weigh up to 8 lbs.

2. Cottontail Rabbit: Active in the early morning and late evening

3.Gopher, Kangaroo Rat, Wood Rat, Pocket Mouse, Grasshopper Mouse: Active at night.

4. Ground Squirrel, Tree Squirrel, Chipmunk: Out during the day.

5. Porcupine: Singe the quills, then skin and roast or boil.

6. Muskrat, Beaver: Beaver tail is especially delicious, broil it on a stick then remove the skin.

7. Skunk: Skin carefully, the meat is excellent. Active at night.

8. Fox, Coyote, Bear.

9. Bobcat, Wildcat, Mountain Lion.

10. Dove: Usually found near habitation and water.
11. Deer, Elk: Keep hair off meat.

Elta Alder

OTHER EDIBLES

1. Quail, Grouse, Pheasant

2. Ducks, and other water Fowl

3. Owls, Hawks, Crows, Wrens and various other small birds.

4.Woodpeckers

5. Vultures and Eagles.

6. Birds Eggs: All are edible.

7. Fish.

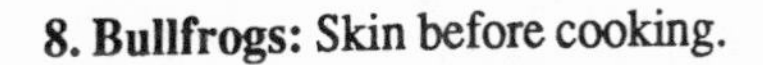

8. Bullfrogs: Skin before cooking.

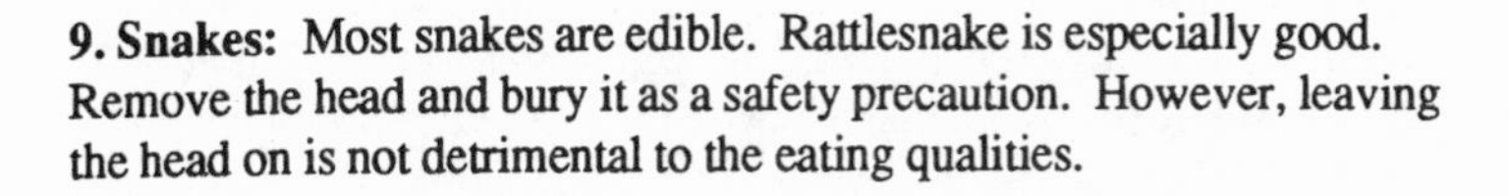

9. Snakes: Most snakes are edible. Rattlesnake is especially good. Remove the head and bury it as a safety precaution. However, leaving the head on is not detrimental to the eating qualities.

10. Lizards.: All believed edible including the poisonous Gila Monster.

11. Lubber Grasshopper: 2 1/2 inches long. Should be cooked.

12. Grubs: Found under bark, in rotten wood or in the ground. Boil or fry.

13. Hairless Caterpillars: Hairy ones may be poisonous

Preserving Surplus Meat

Sand Dried Meat-- is stripped similar to jerky, then wiped dry, and buried, unsalted in dry sand about 6 inches deep. If kept dry will keep for several years. Eat dry or soak and cook.

Smoke Drying--is also simple. Build a lattice about 3 feet above a slow burning fire, lay 1/4 " thick strips of meat on the lattice. Smoke until the meat becomes brittle. Do not let the fire become so hot that the meat cooks or draws juices-the smoke does the trick. Do not use pitchy or oily woods as they will flavor the meat.

Deer Roast

Select a roast, cover it with chopped onions (aprox. 1 cup), 2 or 3 bay leaves, salt and pepper, 1/2 t. ground cloves, and 1/2 t. allspice. Pour vinegar (just enough to cover the meat) in a covered dish and marinate for 12 hours or more. Take the roast out of the pan and fry it in hot oil until brown. Put the roast back in the marinade and cover with a lid. Bake for 2 hours or until tender. Slice and serve.

Elta Alder

Baked Rabbit or Squirrel

Soak the meat overnight in salt water and 2 C vinegar to 2 C water. Stew or fry in the same way you would chicken. To bake a squirrel or rabbit the hair must be removed by skinning it or singeing it in a fire. Clean and gut the animal, and rub it inside and out with oil. Salt and pepper. The Indians rubbed the animal with wood ashes until it turned white, then they baked it before an open fire. You can also bake it in the oven. Save all the drippings to make gravy.

Elta Alder

Quail Casserole

3 Quails cut up
1/2 C butter
1/4 C dried onion
6 C celery
1/2 t. cornstarch
1/2 C chicken broth

Saute quail in butter, then remove the meat and saute the onions, and celery in the same butter. Add cornstarch dissolved in bouillon and cook until thickened. Place the quail in a baking dish and pour the sauce over it. Bake in a 350° oven for 15 minutes.

Elta Alder

Fish (barbecue style)

Cut up fish. Roast over a fire on sticks, turning often until fish no longer drips.

Elta Alder

Quail or Small Birds

Clean the bird and roast on a stick over the fire until brown. Place in a pot of water and boil until well done. Thicken the broth , salt and pepper to taste. Serve over rice or noodles.

Elta Alder

Bird on Toast

3 small birds
1 1/2 t. water
3-5 slices of toast
1 1/2 t. butter
juice from 1 lemon
3 slices of bacon

Clean the birds. Butter the inside of each bird, salt and pepper. Wrap a slice of bacon around each bird. Roast in a baking dish with butter for 20-30 minutes. Serve over toast with gravy made from the drippings, a little butter and the juice of 1 lemon.

Elta Alder

Index

Products Available Through Mail Order:

Books

Cookin' with Home Storage	$14.95
Food Storage 101 Where Do I Begin?	$11.95
Cookin' with Dried Eggs	$ 6.50
Cookin' with Powdered Milk	$ 8.50
Cookin' with Rice and Beans	$11.95
Cookin' with Kids in the Kitchen	$11.95

Upcoming Books Available Soon

Cookin' with Wheat and Other Grains
Cookin' with Dehydrated Foods
Copy Cat Cookin' (Make your own Boxed mixes, like Hamburger Helper, Rice-a-Roni, Oatmeal packets, and, lots, more)
Great Grandma' s' Recipes, Remedies, & Washday Hints

Dehydrated Foods Available

If you can not find dehydrated foods including all varieties of rice and beans locally, you can write to me for sources and price lists available through mail order.

All foods are packaged in gallon (#10 can) containers or 5 gallon buckets.

Order Form

"Cookin' With Home Storage"

To order, fill out order form and send to:

"Cookin' With Home Storage"
P.O. Box 44
Manti, Utah 84642

Phone orders: Peggy Layton (435) 835-0311

Name:______________________________

Address:____________________________

City:__________________ **State:** ______

Zip:___________ **Phone:**______________

Please send me_____copies of "Cookin' with Home Storage" @ $14.95 + $2.25 tax and shipping. Total $17.20. For each additional book deduct $1.00 per book.

(group rates available)

Starters For Cheese:

Rennet tablets, Buttermilk freeze dried culture, Yogurt freeze dried culture, Cheese coloring

Send for current prices of these and other food storage products to:

Peggy Layton
P.O. Box 44
Manti, UT 84642
(435) 835-0311

Name:____

Address:______________________________________

City:______________________________**State:**______

Zip:________________**Phone:**____________________

Please send me the following:

____

____

____

____

____

Or list selections on a seperate sheet of paper. Enclose tax (Utah residents add 6.25%) and $1.75 shipping for the first book and $1.00 for each additional book.

Total Amount enclosed $____________.

Group Rates & Wholesale Pricing Available

NOTES

NOTES